WHAT EVERY ENGINEER SHOULD KNOW ABOUT COMPUTATIONAL TECHNIQUES OF FINITE ELEMENT ANALYSIS

Second Edition

WHAT EVERY ENGINEER SHOULD KNOW ABOUT COMPUTATIONAL TECHNIQUES OF FINITE ELEMENT ANALYSIS

Second Edition

LOUIS KOMZSIK

CRC Press
Taylor & Francis Group
Boca Raton   London   New York

CRC Press is an imprint of the
Taylor & Francis Group, an **informa** business

CRC Press
Taylor & Francis Group
6000 Broken Sound Parkway NW, Suite 300
Boca Raton, FL 33487-2742

© 2009 by Taylor & Francis Group, LLC
CRC Press is an imprint of Taylor & Francis Group, an Informa business

No claim to original U.S. Government works
Printed in the United States of America on acid-free paper
10 9 8 7 6 5 4 3 2 1

International Standard Book Number-13: 978-1-4398-0294-6 (Hardcover)

This book contains information obtained from authentic and highly regarded sources. Reasonable efforts have been made to publish reliable data and information, but the author and publisher cannot assume responsibility for the validity of all materials or the consequences of their use. The authors and publishers have attempted to trace the copyright holders of all material reproduced in this publication and apologize to copyright holders if permission to publish in this form has not been obtained. If any copyright material has not been acknowledged please write and let us know so we may rectify in any future reprint.

Library of Congress Cataloging-in-Publication Data

Komzsik, Louis.
    What every engineer should know about computational techniques of finite element analysis / Louis Komzsik. -- 2nd ed.
        p. cm. -- (What every engineer should know ; 38)
    Includes bibliographical references and index.
    ISBN 978-1-4398-0294-6 (hardcover : alk. paper)
    1. Finite element method. I. Title. II. Series.

TA347.F5K665 2009
620.001'51825--dc22   2009004459

Visit the Taylor & Francis Web site at
http://www.taylorandfrancis.com

and the CRC Press Web site at
http://www.crcpress.com

To my son, Victor

Contents

viii

III   Engineering Solution Computations 235

Preface to the second edition

I am grateful to Taylor & Francis, in particular to Nora Konopka, publisher, for the opportunity to revise this book after five years in print, and for her enthusiastic support of the first edition. This made the book available to a wide range of students and practicing engineers fulfilling my original intentions. My sincere thanks are also due to Amy Blalock, project coordinator, and Michele Dimont, project editor, at Taylor & Francis.

Mike Gockel, my colleague of many years, now retired, was again instrumental in clarifying some of the presentation, and he deserves my repeated gratitude. I would like to thank Professor Duc Nguyen for his proofreading of the extensions of this edition. His use of the first edition in his teaching provided me with valuable feedback and confirmation of the approach of the book.

A half a decade passed since the original writing of the first edition and this edition contains numerous noteworthy technical extensions. In Part I the finite element chapter now contains a brief introduction to quadratic finite element shape functions (1.8). Also in Part I, the geometry modeling chapter has been extended with three sections (2.3, 2.4 and 2.5) to discuss the B-spline technology that has become the de facto industry standard. Several new sections were added to address reader requested topics, such as supporting the rigid body motion (4.6), the method of augmenting constraints (4.7) and a discussion on detecting and eliminating massless mechanisms (5.5).

Still in Part I, a new Chapter 6 describes a significant application trend of the past years: the use of the technology to couple multiple physical phenomena. This includes a more detailed description of the fluid-structure interaction application, a hexahedral finite element, as well as a structural-acoustics case study.

In Part II, a new section (7.7) addressing iterative solutions of linear systems and specifically the method of conjugate gradients, was also recommended by readers of the first edition. Also in Part II, a new Chapter 10 is dedicated to complex spectral computations, a topic briefly mentioned but not elaborated on in the first edition. The rotor dynamic application topic and related case study examples round up this new chapter.

In Part III, the modal solution chapter has been extended with a new section

(13.6) describing modal energies and contributions. A new section (14.6) in the transient response analysis chapter discusses the state-space formulation. The frequency domain analysis chapter has been enhanced with a new section (15.5) on enforced motion computations. Finally, the nonlinear chapter received a new section (16.2) describing geometric nonlinearity computations in some detail.

The application focus has also significantly expanded during the years since the publication of the first edition and one of the goals of this edition was to reflect these changes. The updated case study sections' (2.8, 7.6, 9.8, 10.8, 12.6, 14.5, 15.4 and 18.5) state-of-the-art application results demonstrate the tremendously increased computational complexity.

The final goal of this edition was to correct some of the typing mistakes and technical misstatements of the first edition, which were pointed out to me by readers. While they kindly stated that those were not limiting the usefulness of the book, I exercised extreme caution to make this edition as error free and clear as possible.

Louis Komzsik
2009

The model in the cover art is courtesy of Pilates Aircraft Corporation, Stans, Switzerland. It depicts the tail wing vibrations of a PC-21 aircraft, computed by utilizing the techniques described in this book.

Preface to the first edition

The method of finite elements has become a dominant tool of engineering analysis in a large variety of industries and sciences, especially in mechanical and aerospace engineering. In this role, the method enables the engineer or scientist to solve a physical problem or analyze a process. There is, however, significant computational work - in several distinct phases - involved in the solution of a physical problem with the finite element method. The emphasis of this book is on the computational techniques of this complete process from the physical problem to the computed solution.

In the first phase the physical problem is described in mathematical form, most of the time by a boundary value problem of some sort. At the same time the geometry of the physical problem is also approximated by computational geometry techniques resulting in the finite element model. Applying boundary conditions and various constraints to the finite element model results in a numerically solvable form. The first part of the book addresses these topics.

In the second phase of operations the numerical model is reduced to a computationally more efficient form via various spectral representations. Today finite element problems are extremely large in industrial applications, therefore, this is an important step. The subject of the second part of the book is the reduction techniques to reach an efficiently solvable computational model.

Finally, the solution of the engineering problem is obtained with specific computational techniques. Both time and frequency domain solutions are used in practice. Advanced computations addressing nonlinearity and optimization may also be applied. The third part of the book deals with these topics as well as the representation of the computed results.

The book is intended to be a concise, self-contained reference for the topic and aimed at practicing engineers who put the finite element technique to practical use. It may be the subject of specific interest to users of commercial finite element analysis products, as those products execute most of these computational techniques in various forms. Graduate students of finite element techniques in any discipline could benefit from using the book as well.

The material comes from my three decades of activity in the shipbuilding, aerospace and automobile industries, during which I used many of these techniques. I have also personally implemented some of these techniques into various versions of NASTRAN[1], the world's leading finite element software.

Finally, I have also encountered many students during my years of teaching whose understanding of these computations would have been significantly better with such a book.

Louis Komzsik
2004

[1] - NASTRAN is a registered trademark of the National Aeronautics and Space Administration

Acknowledgments

I appreciate Mr. Mike Gockel's (MSC Software Corporation, retired) technical evaluation of the manuscript and his important recommendations, especially those related to the techniques of Chapters 4 and 5.

I would also like to thank Dr. Al Danial (Northrop-Grumman Corporation) for his repeated and very careful proofreading of the entire manuscipt. His clarifying comments representing the application engineer's perspective have significantly contributed to the readability of the book.

Professor Barna Szabo (Washington University, St. Louis) deserves credit for his valuable corrections and insightful advice through several revisions of the book. His professional influence in the subject area has reached a wide range of engineers and analysts, including me.

Many thanks are also due to Mrs. Lori Lampert (MSC Software Corporation) for her expertise and patience in producing figures from my hand-drawings.

I also value the professional contribution of the publication staff at Taylor and Francis Group. My sincere thanks to Nora Konopka, publisher, Helena Redshaw, manager and editor Richard Tressider. They all deserve significant credit in the final outcome.

Louis Komzsik
2004

Part I

Numerical Model Generation

1

Finite Element Analysis

The goal of this chapter is to introduce the reader to finite element analysis which is the basis for the discussion of the computational methods in the remainder of the book. This chapter first focuses on the computational fundamentals of the method in connection with a simple boundary value problem. These fundamentals will be expanded with the derivation of a practical finite element and further when dealing with the application of the technique for mechanical systems in Chapter 3.

1.1 Solution of boundary value problems

The method of using finite elements for the solution of boundary value problems has almost a century of history. The pioneering paper by Ritz [8] has laid the foundation for this technology. The most widely used practical technique, however, is Galerkin's method [3].

The difference between the Ritz method and that of Galerkin's is in the fact that the first addresses the variational form of the boundary value problem. Galerkin's method minimizes the residual of the differential equation integrated over the domain with a weight function, hence it is also called the method of weighted residuals.

This difference lends more generality and computational convenience to Galerkin's method. Let us consider a linear differential equation in two variables on a simple domain D:

$$L(q(x,y)) = 0, (x,y) \in D,$$

and apply Dirichlet boundary conditions on the boundary B

$$q(x,y) = 0, (x,y) \in B.$$

Galerkin's method is based on the Ritz's approximate solution idea and constructs the approximate solution as

$$\bar{q}(x,y) = q_1 N_1 + q_2 N_2 + \ldots + q_n N_n,$$

where the q_i are the yet unknown solution values at discrete points in the domain (the node points of the finite element mesh) and

$$N_i, \quad i = 1, ..n,$$

is the set of the finite element shape functions to be derived shortly. In this case, of course there is a residual of the differential equation

$$L(\bar{q}) \neq 0.$$

Galerkin proposed using the shape functions of the approximate solution also as the weights, and requires that the integral of the so weighted residual vanish.

$$\int\int_D L(\bar{q}) N_j(x,y) dx dy = 0; j = 1, 2, \ldots, n.$$

This yields a system for the solution of the coefficients as

$$\int\int_D L(\sum_{i=1}^{n} q_i N_i(x,y)) N_j(x,y) dx dy = 0; j = 1, 2, \ldots, n.$$

This is a linear system and produces the unknown values of q_i.

Let us now consider the deformation of an elastic membrane loaded by a distributed force of $f(x,y)$ shown in Figure 1.1. The mathematical model is the well-known Poisson's equation.

$$-\frac{\partial^2 q}{\partial x^2} - \frac{\partial^2 q}{\partial y^2} = f(x,y),$$

where $q(x,y)$ is the vertical displacement of the membrane at (x,y) and $f(x,y)$ is the distributed load on the surface of the membrane. Assume the membrane occupies the D domain in the $x-y$ plane with a boundary B. We assume that the membrane is clamped manifested by a Dirichlet boundary condition. It should be noted that in practical problems the boundary is not necessarily as smooth as shown on the Figure 1.1, in fact it is usually only piecewise analytic.

Let us now apply Galerkin's method to this problem.

$$\int\int_D -(\frac{\partial^2 \bar{q}}{\partial x^2} + \frac{\partial^2 \bar{q}}{\partial y^2} + f(x,y)) N_j dx dy = 0, j = 1, \ldots, n.$$

Substituting the approximate solution yields

$$\int\int_D -(\sum_{i=1}^{n} q_i \frac{\partial^2 N_i}{\partial x^2} + \sum_{i=1}^{n} q_i \frac{\partial^2 N_i}{\partial y^2} + f(x,y)) N_j dx dy = 0, j = 1, \ldots, n.$$

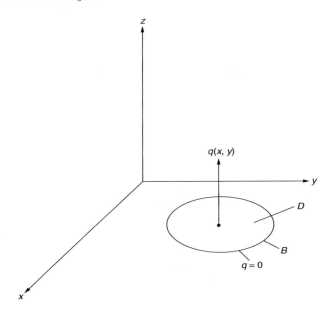

FIGURE 1.1 Membrane model

The left hand side terms may be integrated by parts and after employing the boundary condition they simplify as

$$\int\int_D -(\frac{\partial^2 N_i}{\partial x^2} + \frac{\partial^2 N_i}{\partial y^2})N_j dxdy = \int\int_D (\frac{\partial N_i}{\partial x}\frac{\partial N_j}{\partial x} + \frac{\partial N_i}{\partial y}\frac{\partial N_j}{\partial y})dxdy.$$

Substituting and regrouping yields

$$\int\int_D (\sum_{i=1}^n q_i \frac{\partial N_i}{\partial x}\frac{\partial N_j}{\partial x} + \sum_{i=1}^n q_i \frac{\partial N_i}{\partial y}\frac{\partial N_j}{\partial y} - f(x,y)N_j)dxdy = 0, j = 1, \ldots, n.$$

Unrolling the sums and reordering we get the Galerkin equations:

$$\int\int ((q_1 \frac{\partial N_1}{\partial x} + \ldots + q_n \frac{\partial N_n}{\partial x})\frac{\partial N_j}{\partial x} + (q_1 \frac{\partial N_1}{\partial y} + \ldots + q_n \frac{\partial N_n}{\partial y})\frac{\partial N_j}{\partial y})dxdy =$$

$$\int\int f(x,y)N_j dxdy$$

for $j = 1, .., n$. Introducing the notation

$$K_{ij} = K_{ji} = \int\int (\frac{\partial N_i}{\partial x}\frac{\partial N_j}{\partial x} + \frac{\partial N_i}{\partial y}\frac{\partial N_j}{\partial y})dxdy$$

and

$$F_j = \int \int (f(x,y)N_j)dxdy$$

the Galerkin equations may be written as a matrix equation

$$Kq = F.$$

The system matrix is

$$K = \begin{bmatrix} K_{1,1} & K_{1,2} & \dots & K_{1,n} \\ K_{2,1} & K_{2,2} & \dots & K_{2,n} \\ \dots & \dots & \dots & \dots \\ K_{n,1} & K_{n,2} & \dots & K_{n,n} \end{bmatrix},$$

with solution vector of

$$q = \begin{bmatrix} q_1 \\ q_2 \\ \dots \\ q_n \end{bmatrix},$$

and right hand side vector of

$$F = \begin{bmatrix} F_1 \\ F_2 \\ \dots \\ F_n \end{bmatrix}.$$

The assembly process is addressed in more detail in Section 1.4 after introducing the shape functions. The K matrix is usually very sparse as many K_{ij} become zero. This equation is known as the linear static analysis problem, where K is called the stiffness matrix, F is the load vector and q is the vector of displacements, the solution of Poisson's equation. Other differential equations could lead to similar form as demonstrated in, for example [2].

The concept, therefore, is generally contributing to its wide-spread application success. For the mathematical theory see [6]; the matrix algebraic foundation is thoroughly discussed in [7]. More details may be obtained from the now classic text of [11].

1.2   Finite element shape functions

To interpolate inside the elements piecewise polynomials are usually used. For example a triangular discretization of a two dimensional domain may be approximated by bilinear interpolation functions of form

$$q(x, y) = a + bx + cy.$$

In order to find the coefficients let us consider the triangular region (element) of the $x - y$ plane in a specifically located local coordinate system and the notation shown in Figure 1.2.

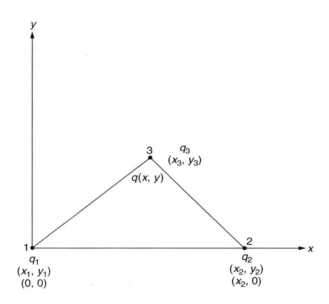

FIGURE 1.2 Local coordinates of triangular element

The usage of a local coordinate system in Figure 1.2 does not limit the generality of the following discussion. The arrangement can always be achieved by appropriate coordinate transformations on a generally located triangle. Using the notation and assignments on Figure 1.2 and by evaluating at each node of the triangle

$$q_e = \begin{bmatrix} q_1 \\ q_2 \\ q_3 \end{bmatrix} = \begin{bmatrix} 1 & 0 & 0 \\ 1 & x_2 & 0 \\ 1 & x_3 & y_3 \end{bmatrix} \begin{bmatrix} a \\ b \\ c \end{bmatrix}.$$

The triangular system of equations is easily solved for the unknown coefficients as

$$\begin{bmatrix} a \\ b \\ c \end{bmatrix} = \begin{bmatrix} 1 & 0 & 0 \\ -\dfrac{1}{x_2} & \dfrac{1}{x_2} & 0 \\ \dfrac{x_3 - x_2}{x_2 y_3} & \dfrac{-x_3}{x_2 y_3} & \dfrac{1}{y_3} \end{bmatrix} \begin{bmatrix} q_1 \\ q_2 \\ q_3 \end{bmatrix}.$$

By back-substituting into the approximation equation we get

$$q(x, y) = N \begin{bmatrix} q_1 \\ q_2 \\ q_3 \end{bmatrix} = \begin{bmatrix} N_1 & N_2 & N_3 \end{bmatrix} \begin{bmatrix} q_1 \\ q_2 \\ q_3 \end{bmatrix}.$$

Here N contains the N_1, N_2, N_3 shape functions (more precisely the traces of shape functions inside an element). With these we are now able to describe the relationship between the solution value inside an element in terms of the solutions at the corner node points

$$q(x, y) = N_1 q_1 + N_2 q_2 + N_3 q_3.$$

The values of N_i are

$$N_1 = 1 - \frac{1}{x_2} x + \frac{x_3 - x_2}{x_2 y_3} y,$$

$$N_2 = \frac{1}{x_2} x - \frac{x_3}{x_2 y_3} y,$$

and

$$N_3 = \frac{1}{y_3} y.$$

These clearly depend on the coordinates of the corner node of the particular triangular element of the domain. It is easy to see that at every node only one of the shape functions is nonzero. Specifically at node 1: N_2 and N_3 vanish, while $N_1 = 1$. At node 2: $N_2 = 1$, both N_1 and N_3 are zero. Finally at node 3: N_3 takes a value of one and the other two vanish. It is also easy to verify that the

$$N_1 + N_2 + N_3 = 1$$

equation is satisfied.

The nonzero shape functions at a certain node point reduce to zero at the other two nodes, respectively. The interpolations are continuous across the neighboring elements. On an edge between two triangles, the approximation is linear. It is the same when it is approached from either element.

Specifically along the edge between nodes 1 and 2 the shape function N_3 is zero. The shape functions N_1 and N_2 along this edge are the same when calculated from an element on either side of that edge.

Naturally, additional computations are required to reflect to the fact when the triangle is generally located, i.e. none of its sides is collinear with any axes. This issue of local-global coordinate transformations will be discussed shortly.

)

1.3   Finite element basis functions

There is another (sometimes misinterpreted) component of finite element technology, the basis functions. They are sometimes used in place of shape functions by engineers, although as shown below, they are distinctly different. The approximation of

$$q(x, y) = N q_e$$

may also be written as

$$q(x, y) = M c_e$$

where M is the matrix of basis functions and c_e is the vector of basis coefficients.

Clearly for our example

$$M = \begin{bmatrix} 1 & x & y \end{bmatrix}$$

and

$$c_e = \begin{bmatrix} a \\ b \\ c \end{bmatrix}.$$

The family of basis functions for two-dimensional elements may be written from the terms shown on Table 1.1.

Depending on how the basis functions are chosen, various two-dimensional elements may be derived. Naturally a higher order basis function family requires more node points. For example, a quadratic (order= 2) triangular element, often used in industry, is based on introducing midpoint nodes on each side of the triangle. This enables the use of the following interpolation

TABLE 1.1
Basis function terms for
two-dimensional elements

Order	Terms			
0		1		
1	x		y	
2	x^2	xy	y^2	
3	x^3	x^2y	xy^2	y^3

function

$$q(x, y) = a + bx + cy + dx^2 + ey^2 + fxy$$

in each triangle. The six coefficients are again easily established by a procedure similar to the linear triangular element above. The interpolation across quadratic element boundaries is also continuous, however, now it is parabolic along an edge. Nevertheless, the parabola produced by the neighboring elements is the same from both sides. Quadratic finite elements will be discussed in section 1.8.

For a first order rectangular element the interpolation may be of the form

$$q(x, y) = a + bx + cy + dxy.$$

In this case, all the first-order basis functions were used as well as one component of the second-order basis function family. We will derive a practical rectangular element in Section 1.7. Similarly a second-order (eight noded) rectangular element is approximated as

$$q(x, y) = a + bx + cy + dxy + ex^2 + fx^2y + gxy^2 + hy^2.$$

This is again the use of the complete 2nd order family plus two components of the 3rd order family to accommodate additional node points. The latter are usually located on the midpoints of each side, as they were on the quadratic triangle.

For a three-dimensional domain, the four noded tetrahedron is one of the most commonly used finite elements. The interpolation inside a tetrahedral element is of form

$$q(x, y, z) = a + bx + cy + dz.$$

The basis function terms for three-dimensional elements is shown in Table 1.2.

Quadratic interpolation of the tetrahedron is also possible; the related element is called the 10-noded tetrahedron. The extra node points are located

TABLE 1.2
Basis function terms for
three-dimensional elements

Order		Terms			
0		1			
1		x y	z		
2	x^2 xy y^2		xz	yz	z^2
3	x^3 ...	xyz	...	y^3	z^3

on the midpoints of the edges.

$$q(x, y, z) = a + bx + cy + dz + ex^2 + fxy + gy^2 + hxz + iyz + jz^2.$$

The third-order three-dimensional basis function family introduces another 10 terms, some of them are shown on Table 1.2.

Finally, additional volume elements are also frequently used. The hexahedron is one of the most widely accepted. Its first order version consists of eight node points at the corners of the hexahedron and therefore, it is defined with specifically chosen basis functions as

$$q(x, y, z) = a + bx + cy + dz + exy + fxz + gyz + hxyz.$$

The quadratic hexahedral element consists of 20 nodes, the eight corner nodes and the 12 mid points on the edges. A 3rd order hexahedral element with 27 nodes is also used, albeit not widely. The additional seven nodes come from the mid-point of the six faces and from the center of the volume.

Finally, higher order polynomial (p-version) elements are also used in the industry. These elements introduce side shape functions in addition to the nodal shape functions mentioned earlier. The side shape functions, as their name indicates, are assigned to the sides of the elements. They are formulated in terms of some orthogonal, most often Legendre, polynomial of order p, hence the name. There are clearly advantages in computational accuracy when applying such elements. On the other hand, they introduce extra computational costs, so they are mainly used in specific applications and not generally. The method and some applications are described in detail in the book of the pioneering authors of the technique [9].

The gradual widening of the finite element technology may be assessed by reviewing the early articles of [10] and [1], as well as from the reference of the first general purpose and still premier finite element analysis tool [4].

1.4   Assembly of finite element matrices

The repeated application of general triangles may be used to cover the D planar domain as shown in Figure 1.3. The process is called meshing. The points inside the domain and on the boundary are the node points. They span the finite element mesh.   There may be small gaps between the boundary and the

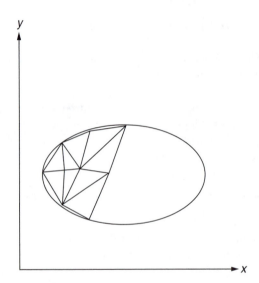

FIGURE 1.3 Meshing the membrane model

sides of the triangles adjacent to the boundary. This issue contributes to the approximation error of the finite element method. The gaps may be filled by progressively smaller elements or those triangles may be replaced by triangles with curved edges. Nevertheless, all the elemental matrices contribute to the global finite element matrices and the process of computing these contributions is the finite element matrix assembly process.

One way to view the assembly of the K matrix is by way of the shape functions. For the triangular element discussed in the last section a shape function associated with a node describes a plane going through the other two nodes and having a height of unity above the associated node. On the other hand, in an adjacent element the shape function associated with the same node describes another plane, and so on. In general, a shape function N_i will define a pyramid over node i.

This geometric interpretation explains the sparsity of the K matrix. Only those $N_i N_j$ products will exist, and in turn produce a K_{ij} entry in the K matrix, where the two pyramids of N_i and N_j overlap.

A computationally more practical method is based on summing up the energy contributions from each element to the global matrix. The strain energy (a component of the potential energy) of a certain element is

$$E_e = \frac{1}{2} \int \int [(\frac{\partial q}{\partial x})^2 + (\frac{\partial q}{\partial y})^2] dx dy.$$

Introducing the strain vector

$$\epsilon = \begin{bmatrix} \frac{\partial q}{\partial x} \\ \frac{\partial q}{\partial y} \end{bmatrix}.$$

the strain energy of the element is

$$E_e = \frac{1}{2} \int \int \epsilon^T \epsilon\, dx dy.$$

Considering our simple triangular element, differentiating and using a matrix notation yields

$$\epsilon = \begin{bmatrix} \frac{\partial q}{\partial x} \\ \frac{\partial q}{\partial y} \end{bmatrix} = \begin{bmatrix} b \\ c \end{bmatrix} = \begin{bmatrix} -\frac{1}{x_2} & \frac{1}{x_2} & 0 \\ \frac{x_3 - x_2}{x_2 y_3} & \frac{-x_3}{x_2 y_3} & \frac{1}{y_3} \end{bmatrix} \begin{bmatrix} q_1 \\ q_2 \\ q_3 \end{bmatrix} = B q_e,$$

where

$$q_e = \begin{bmatrix} q_1 \\ q_2 \\ q_3 \end{bmatrix}.$$

In the above, B is commonly called the strain-displacement matrix. The $\frac{\partial q}{\partial x}$ and $\frac{\partial q}{\partial y}$ terms are the strain components of our element, in essence the rate of change of the deformation of the element in the coordinate directions. The B matrix relates the strains to the nodal displacements on the right, hence the name.

Note, that the structure of B depends on the physical model, in our case having only one degree of freedom per node point for the membrane element. Elements representing other physical phenomena, for example, triangles having two in-plane degrees of freedom per node point, have different B matrix, as they have more possible strain components. This issue will be addressed in more detail in Section 1.7 and in Chapter 3. Here we stay on a mathematical focus.

With this the element energy contribution is

$$E_e = \frac{1}{2} \int \int q_e^T B^T B q_e \, dx dy.$$

Since the node point coordinates are constant with respect to the integration we may write

$$E_e = \frac{1}{2} q_e^T \left(\int \int B^T B \, dx dy \right) q_e = \frac{1}{2} q_e^T k_e q_e.$$

Here k_e is the element matrix whose entries depend only on the shape of the element. If our element is the element described by nodes 1, 2 and 3, then the terms in k_e contribute to the terms of the 1st, 2nd and 3rd columns and rows of the global K matrix. The actual integration for computing k_e is addressed in the next section.

Let us assume that another element is adjacent to the 2-3 edge, its other node being 4. Then by similar arguments, the 2nd element's matrix terms (depending on that particular element's shape) will contribute to the 2nd, 3rd and 4th columns and rows of the global matrix. This process is continued for all the elements contained in the finite element mesh.

Note, that in the case of quadratic or quadrilateral shape elements the actual element matrices are again of different sizes. This fact is due to the different number of node points describing the element geometry. Nevertheless, the matrix generation and assembly process is conceptually the same.

Furthermore, in the case of three-dimensional elements the energy formulation is even more complex. These issues will be discussed in more detail in Chapter 3.

1.5   Element matrix generation

Let us now focus on calculating the element matrix integrals. Since for our model B is constant (function of only the coordinates of the node points of the element), this may be simplified to

$$k_e = B^T B \int\int dxdy = B^T B A_e,$$

where A_e is the surface area of the element as

$$A_e = \int\int dxdy.$$

In order to evaluate this integral, the element is usually represented in parametric coordinates. Let us consider again the local coordinates of the triangular element, now shown in Figure 1.4 with two specific coordinate axes representing the parametric system. The axis (coincident with the local x axis) going through node points 1 and 2 is the first parametric axis u. Define the other axis going from node 1 through node 3 as v. If we define the $(0,0)$ parametric location to be node 1, the $(1,0)$ to be node 2 and the $(0,1)$ to be node 3, then the parametric transformation is of form

$$u = \frac{1}{x_2}x - \frac{x_3}{x_2 y_3}y$$

and

$$v = \frac{1}{y_3}y.$$

Here we took advantage of the local coordinates of the nodes as shown on Figure 1.2. Note, that this transformation may also be written

$$u = N_2$$

and

$$v = N_3.$$

Furthermore, the points inside one element may be written as

$$x = N_1 x_1 + N_2 x_2 + N_3 x_3,$$

and

$$y = N_1 y_1 + N_2 y_2 + N_3 y_3.$$

Since we describe the coordinates of a point inside an element with the same shape functions that were used to approximate the displacement field, this is called an iso-parametric representation and our element is called an iso-parametric element. Applying the local coordinates of our element of Figure

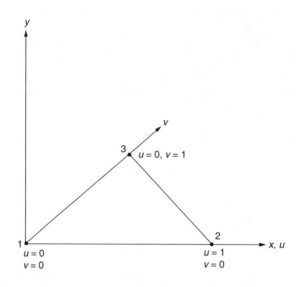

FIGURE 1.4 Parametric coordinates of triangular element

1.2 yields

$$x = x_2 u + x_3 v$$

and

$$y = y_3 v.$$

The integral with this parameterization is

$$\int \int dx dy = \int \int det[\frac{\partial(x,y)}{\partial(u,v)}]dudv.$$

Here the Jacobian matrix

$$\frac{\partial(x,y)}{\partial(u,v)} = \begin{bmatrix} \frac{\partial x}{\partial u} & \frac{\partial x}{\partial v} \\ \frac{\partial y}{\partial u} & \frac{\partial y}{\partial v} \end{bmatrix} = \begin{bmatrix} x_2 & x_3 \\ 0 & y_3 \end{bmatrix}.$$

With this result

$$A_e = x_2 y_3 \int \int dudv.$$

In practice the parametric integral for each element is executed numerically, most commonly via Gaussian numerical integration, quadrature for two dimensions and cubature for three dimensions. Note, that this is in essence a reduction type computation, main focus of Part II, as opposed to the analytic integration over the continuum domain.

Gaussian numerical integration has become the industry standard tool for integration of the element matrices by virtue of its higher accuracy than the Newton-Cotes type methods such as Simpson's. In general an integral over a specific continuous interval is approximated by a sum of weighted function values at some specific locations.

$$\int_{-1}^{1} f(t)dt = \Sigma_{i=1}^{n} c_i f(t_i).$$

Here n is the number of integration points used. The specific sampling locations are the zeroes of the n-th Legendre polynomial:

$$t_i : P_n(t) = 0$$

and the recursive definition of Legendre polynomials is

$$(k+1)P_{k+1}(t) = (2k+1)tP_k(t) - kP_{k-1}(t).$$

Starting from $P_0(t) = 1$ and $P_1(t) = t$ the recurrence form produces

$$P_2(t) = \frac{1}{2}(3t^2 - 1),$$

$$P_3(t) = \frac{1}{2}(5t^3 - 3t),$$

and so on. The c_i weights are computed as

$$c_i = \int_{-1}^{1} L_{n-1,i}(t)dt$$

where

$$L_{n,i} = \prod_{j=1,j\neq i}^{n} \frac{t - t_j}{t_i - t_j}$$

is the i-th n-th order Lagrange polynomial with roots of the Legendre polynomials described above. For the most commonly occurring cases Table 1.3 shows the values of c_i and t_i.

Now integrating over the parametric domain of our element, the integral has the following boundaries

$$\int_{u=0}^{1} \int_{v=0}^{1-u} dv\,du.$$

This is clear when looking at Figure 1.4. One needs to transform above integral boundaries to the standard $[-1,1]$ interval required by the Gaussian

TABLE 1.3
Gauss weights and
locations

n	t_i	c_i
1	0	2
2	$\frac{1}{\sqrt{3}}, -\frac{1}{\sqrt{3}}$	1, 1
3	$\sqrt{\frac{3}{5}}, 0, -\sqrt{\frac{3}{5}}$	$\frac{5}{9}, \frac{8}{9}, \frac{5}{9}$

numerical integration. This may be done with the transformation

$$v = \frac{1-u}{2} + \frac{1-u}{2}r,$$

and

$$dv = \frac{1-u}{2}dr,$$

as well as

$$u = \frac{1}{2} + \frac{1}{2}s,$$

and

$$du = \frac{1}{2}ds.$$

The transformed integral amenable to Gauss quadrature is

$$\int_{s=-1}^{1} \frac{1}{2} \int_{r=-1}^{1} (\frac{1}{4} - \frac{1}{4}s)drds.$$

Using the 1-point Gauss formula this is

$$\frac{1}{2}2(\frac{1}{4}2) = \frac{1}{2}.$$

With this the surface area of the element is

$$A_e = \frac{x_2y_3}{2},$$

which agrees with the geometric computation based on the triangle's local
coordinates. This is a rather roundabout way of computing the area of a
triangle. Note, however, that the discussion here is aimed at introducing generally applicable principles.

Naturally, there is a wealth of element types used in various industries.
Even for the simple triangular geometry there are other formulations. The
extensions are in both the number of node points describing the triangular
element as well as in the number of degrees of freedom associated with a node
point.

1.6 Local to global coordinate transformation

When the element matrix assembly issue was addressed earlier, the element matrix had been developed in terms of local (x, y, z) coordinates. In the case of multiple elements, all the elements have their respective local coordinate system chosen on the same principle of the local x axis being collinear with one of the element sides and another one perpendicular.

Thus before assembling any element, the element matrix must be transformed to the global coordinate system common to all the elements. Let us denote the element's local coordinate systems with (x, y, z) and the global coordinate system with (X, Y, Z). The unit direction vectors of the two coordinate systems are related as

$$\begin{bmatrix} i \\ j \\ k \end{bmatrix} = T \begin{bmatrix} I \\ J \\ K \end{bmatrix},$$

where the terms of the transformations are easily obtained from the geometric relation between the local and global systems. Specifically

$$T = \begin{bmatrix} t_{11} & t_{12} & t_{13} \\ t_{21} & t_{22} & t_{23} \\ t_{31} & t_{32} & t_{33} \end{bmatrix},$$

where the t_{mn} term is the cosine of the angle between the mth local coordinate axis and the nth global coordinate axis. The same transformation is applicable to the nodal degrees of freedom of any element

$$\begin{bmatrix} q_x \\ q_y \\ q_z \end{bmatrix} = T \begin{bmatrix} q_X \\ q_Y \\ q_Z \end{bmatrix}.$$

Hence, the element displacements in the two systems are related as

$$q_e = G^{lg} q_e^g,$$

where the upper left and the lower right 3×3 blocks of the 6×6 G^{lg} matrix are the same as the T matrix, the other blocks are zero. The q_e^g notation refers to the element displacements in the global coordinate system.

Considering the element energy contribution

$$E_e = \frac{1}{2} q_e^T k_e q_e$$

and substituting above we get

$$E_e = \frac{1}{2} q_e^{g,T} G^{lg,T} k_e G^{lg} q_e^g$$

or

$$k_e^g = G^{lg,T} k_e G^{lg}.$$

This transformation follows the element matrix generation and precedes the assembly process. Naturally, the solution q_e^g is also represented in global coordinates, which is the subject of the interest of the engineer anyways.

This issue will not be further discussed, the elements introduced later will be generated either in terms of local or global coordinates for simplifying the particular discussion. Commercial finite element analysis systems have specific rules for the definition of local coordinates for various element types.

1.7   A linear quadrilateral finite element

So far we have discussed the rather limited triangular element formulation, mainly to provide a foundation for presenting the integration and assembly computations. We continue this chapter with the discussion of a more practical quadrilateral or rectangular element, but first we focus on the linear case. Quadrilateral elements are the most frequently used elements of industrial finite element analysis when analyzing topologically two-dimensional models, such as the body of an automobile or an airplane fuselage.

Let us place the element in the $x - y$ plane as shown in Figure 1.5, but pose no other restriction on its location.   Based on the principles we developed in connection with the simple triangular element, we introduce shape functions. As we have four nodes in a quadrilateral element, we will have four shape functions, each of whose values vanish at any other node but one. For $j = 1, 2, 3, 4$, we define

$$N_i = \begin{cases} 1 \text{ when } i = j, \\ 0 \text{ when } i \neq j. \end{cases}$$

We create an element parametric coordinate system u, v originated in the interior of the element and having the following definition:

$$N_i = \frac{1}{4}(1 + uu_i)(1 + vv_i).$$

Such a coordinate system is shown in Figure 1.6. The mapping of the general

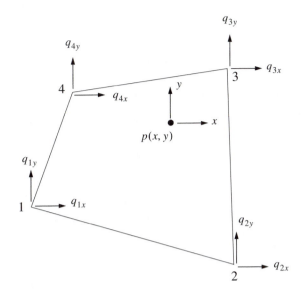

FIGURE 1.5 A planar quadrilateral element

element to the parametric coordinates is the following counterclockwise pattern:

$$(x_1, y_1) \rightarrow (-1, -1),$$
$$(x_2, y_2) \rightarrow (1, -1),$$
$$(x_3, y_3) \rightarrow (1, 1),$$

and

$$(x_4, y_4) \rightarrow (-1, 1).$$

The corresponding four shape functions are:

$$N_1 = \frac{1}{4}(1 - u)(1 - v),$$

$$N_2 = \frac{1}{4}(1 + u)(1 - v),$$

$$N_3 = \frac{1}{4}(1 + u)(1 + v),$$

and

$$N_4 = \frac{1}{4}(1 - u)(1 + v).$$

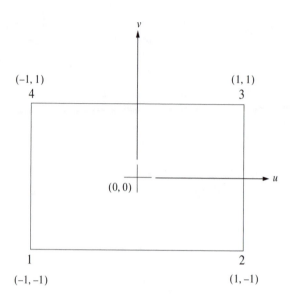

FIGURE 1.6 Parametric coordinates of quadrilateral element

These so-called Lagrangian shape functions will be used for the element formulation. The above selection of the N_i functions obviously again satisfies

$$N_1 + N_2 + N_3 + N_4 = 1.$$

The element deformations, however, will not be vertical to the plane of the element as in the earlier triangular membrane element. This element will have deformation in the plane of the element. Hence, there are eight nodal displacements of the element as

$$q_e = \begin{bmatrix} q_{1x} \\ q_{1y} \\ q_{2x} \\ q_{2y} \\ q_{3x} \\ q_{3y} \\ q_{4x} \\ q_{4y} \end{bmatrix}.$$

The displacement at any location inside this element is approximated with the help of the matrix of shape functions as

$$q(x, y) = N q_e.$$

Since

$$q(x, y) = \begin{bmatrix} q_x(x, y) \\ q_y(x, y) \end{bmatrix}$$

the N matrix of the four shape functions is organized as

$$N = \begin{bmatrix} N_1 & 0 & N_2 & 0 & N_3 & 0 & N_4 & 0 \\ 0 & N_1 & 0 & N_2 & 0 & N_3 & 0 & N_4 \end{bmatrix}.$$

Following the iso-parametric principle also introduced earlier, the location of a point inside the element is approximated again with the same four shape functions as the displacement field:

$$x = N_1 x_1 + N_2 x_2 + N_3 x_3 + N_4 x_4,$$

and

$$y = N_1 y_1 + N_2 y_2 + N_3 y_3 + N_4 y_4.$$

Here x_i, y_i is the location of the i-th node of the element in the x, y directions. Using the shape functions defined above with the element coordinates and by substituting we get

$$x = \frac{1}{4}[(1-u)(1-v)x_1 + (1+u)(1-v)x_2 + (1+u)(1+v)x_3 + (1-u)(1+v)x_4]$$

$$= \frac{1}{4}[(x_1 + x_2 + x_3 + x_4) + u(-x_1 + x_2 + x_3 - x_4) +$$

$$v(-x_1 - x_3 + x_3 + x_4) + uvu(x_1 - x_2 + x_3 - x_4)].$$

Similarly

$$y = \frac{1}{4}[(y_1 + y_2 + y_3 + y_4) + u(-y_1 + y_2 + y_3 - y_4) +$$

$$v(-y_1 - y_3 + y_3 + y_4) + uvu(y_1 - y_2 + y_3 - y_4)].$$

To calculate the element energy and the element matrix, the strain components and the B strain-displacement matrix need to be computed. The element has three constant strains defined from the possible six used in three-dimensional continuum. They are

$$\epsilon = \begin{bmatrix} \frac{\partial q_x}{\partial x} \\ \frac{\partial q_y}{\partial y} \\ \frac{\partial q_x}{\partial y} + \frac{\partial q_y}{\partial x} \end{bmatrix}.$$

We still make an effort here to stay on the mathematical side of the discussion; this will be expanded when modeling a physical phenomenon. Clearly the first two components are the rates of changes in distances between points of the element in the appropriate directions. The third component is a combined rate of change with respect to the other variable in the plane, defining an angular deformation.

The relationship to the nodal displacements is described in matrix form as

$$\epsilon = B q_e.$$

Since the shape functions are given in terms of the parametric coordinates we need again the Jacobian as

$$J = \frac{\partial(x, y)}{\partial(u, v)} = \begin{bmatrix} \frac{\partial x}{\partial u} & \frac{\partial x}{\partial v} \\ \frac{\partial y}{\partial u} & \frac{\partial y}{\partial v} \end{bmatrix} = \frac{1}{4} \begin{bmatrix} j_{11} & j_{12} \\ j_{21} & j_{22} \end{bmatrix}.$$

The terms are

$$j_{11} = -(1-v)x_1 + (1-v)x_2 + (1+v)x_3 - (1+v)x_4,$$

$$j_{12} = -(1-u)x_1 - (1+u)x_2 + (1+u)x_3 + (1-u)x_4,$$

$$j_{21} = -(1-v)y_1 + (1-v)y_2 + (1+v)y_3 - (1+v)y_4,$$

and

$$j_{22} = -(1-u)y_1 - (1+u)y_2 + (1+u)y_3 + (1-u)y_4.$$

Since

$$\begin{bmatrix} \frac{\partial q}{\partial u} \\ \frac{\partial q}{\partial v} \end{bmatrix} = J \begin{bmatrix} \frac{\partial q}{\partial x} \\ \frac{\partial q}{\partial y} \end{bmatrix},$$

the strain components required for the element are

$$\begin{bmatrix} \frac{\partial q_x}{\partial x} \\ \frac{\partial q_x}{\partial y} \end{bmatrix} = J^{-1} \begin{bmatrix} \frac{\partial q_x}{\partial u} \\ \frac{\partial q_x}{\partial v} \end{bmatrix},$$

and

$$\begin{bmatrix} \frac{\partial q_y}{\partial x} \\ \frac{\partial q_y}{\partial y} \end{bmatrix} = J^{-1} \begin{bmatrix} \frac{\partial q_y}{\partial u} \\ \frac{\partial q_y}{\partial v} \end{bmatrix}.$$

Taking advantage of the components of J and using the adjoint-based inverse we compute

$$\epsilon = \begin{bmatrix} \frac{\partial q_x}{\partial x} \\ \frac{\partial q_y}{\partial y} \\ \frac{\partial q_x}{\partial y} + \frac{\partial q_y}{\partial x} \end{bmatrix} = \frac{1}{\det(J)} \begin{bmatrix} j_{22} & -j_{12} & 0 & 0 \\ 0 & 0 & -j_{21} & j_{11} \\ -j_{21} & j_{11} & j_{22} & -j_{12} \end{bmatrix} \begin{bmatrix} \frac{\partial q_x}{\partial u} \\ \frac{\partial q_x}{\partial v} \\ \frac{\partial q_y}{\partial u} \\ \frac{\partial q_y}{\partial v} \end{bmatrix}.$$

From the displacement field approximation equations we obtain

$$
\begin{bmatrix}
\frac{\partial q_x}{\partial u} \\
\frac{\partial q_x}{\partial v} \\
\frac{\partial q_y}{\partial u} \\
\frac{\partial q_y}{\partial v}
\end{bmatrix}
=
$$

$$
\frac{1}{4}
\begin{bmatrix}
-(1-v) & 0 & (1-v) & 0 & (1+v) & 0 & -(1+v) & 0 \\
-(1-u) & 0 & -(1+u) & 0 & (1+u) & 0 & (1-u) & 0 \\
0 & -(1-v) & 0 & (1-v) & 0 & (1+v) & 0 & -(1+v) \\
0 & -(1-u) & 0 & -(1+u) & 0 & (1+u) & 0 & (1-u)
\end{bmatrix}
q_e.
$$

The last two equations produce the B matrix of size 3×12 that is now not constant, it is linear in u and v. Recall that the energy of the element is

$$
E_e = \frac{1}{2} \int \int \epsilon^T \epsilon \, dx dy.
$$

With substitution of $\epsilon = B q_e$ we obtain

$$
E_e = \frac{1}{2} q_e^T \int \int B^T B \, dx dy \, q_e = \frac{1}{2} q_e^T k_e q_e.
$$

The element matrix is

$$
k_e = \int \int B^T B \det\left[\frac{\partial(x,y)}{\partial(u,v)}\right] du dv.
$$

By the fortuitous choice of the parametric coordinate system this integral now is directly amenable to Gaussian quadrature as the limits are $-1, +1$. Introducing

$$
f(u,v) = B^T B \det(J)
$$

the element integral becomes

$$
k_e = \int_{u=-1}^{1} \int_{v=-1}^{1} f(u,v) du dv = \Sigma_{i=1}^n c_i \Sigma_{j=1}^n c_j f(u_i, v_j).
$$

Here u_i, v_j are not the nodal point displacements, but the Gauss point locations (shown as t_i is Table 1.3) in those directions. With applying the two point ($n = 2$) formula

$$
k_e = c_1^2 f(u_1, v_1) + c_1 c_2 f(u_1, v_2) + c_2 c_1 f(u_2, v_1) + c_2^2 f(u_2, v_2)
$$

and c_i are listed in Table 1.3 also.

This concludes the computation techniques of the linear 2-dimensional quadrilateral element. In practice the quadratic version is much preferred and will be described in the following.

1.8   Quadratic finite elements

We view the element in the $x - y$ plane as shown in Figure 1.5, but add nodes on the middle of the sides of the square shown in Figure 1.6 depicting the parametric plane of the element. The locations of these new node points of the quadratic element are:

$$(x_5, y_5) \rightarrow (0, -1),$$

$$(x_6, y_6) \rightarrow (1, 0),$$

$$(x_7, y_7) \rightarrow (0, 1),$$

and

$$(x_8, y_8) \rightarrow (-1, 0).$$

Connecting these points are four interior lines, described by parametric equations as

$$1 - u + v = 0,$$

connecting nodes 5 and 6,

$$1 - u - v = 0,$$

connecting nodes 6 and 7,

$$1 + u - v = 0,$$

connecting nodes 7 and 8, and finally

$$1 + u + v = 0,$$

connecting nodes 8 and 1, completing the loop. For $j = 1, \ldots, 8$ we seek functions N_i that are unit at the $i$the node and vanish at the others:

$$N_i = \begin{cases} 1 \text{ when } i = j, \\ 0 \text{ when } i \neq j. \end{cases}$$

Let us consider for example node 3. N_3 must vanish along the opposite sides of the rectangle

$$u = -1,$$

and

$$v = -1.$$

That will account for nodes $1, 2, 4, 5, 8$. Furthermore it must also vanish at nodes 6 and 7, represented by the line

$$1 - u - v = 0.$$

Hence the form of the corresponding shape function is

$$N_3 = n_c(1+u)(1+v)(1-u-v).$$

where the normalization coefficient n_c for the corner shape functions may be established from the condition of N_3 becoming unit at node 3

$$N_3 = n_c(1+1)(1+1)(1-1-1) = n_c(-4) = 1,$$

yielding

$$n_c = -\frac{1}{4}.$$

This is in part identical to the N_3 shape function of the linear element, apart from the last term. The shape functions corresponding to the corner nodes, based on similar considerations, are of form

$$N_1 = -\frac{1}{4}(1-u)(1-v)(1+u+v),$$

$$N_2 = -\frac{1}{4}(1+u)(1-v)(1-u+v).$$

$$N_3 = -\frac{1}{4}(1+u)(1+v)(1-u-v),$$

and

$$N_4 = -\frac{1}{4}(1-u)(1+v)(1+u-v).$$

To define the shape functions at the mid-points, we consider node 6 first. N_6 must vanish at 3 edges

$$v = 1,$$

$$v = -1,$$

and

$$u = -1.$$

Hence it will be of form

$$N_6 = n_m(1+u)(1+v)(1-v).$$

Substituting the last two terms with the well known algebraic identity, we obtain

$$N_6 = n_m(1+u)(1-v^2).$$

This form now demonstrates the quadratic nature of the element. The normalization constant of the mid-side nodes n_m is established by using the coordinates $(1,0)$ of node 6

$$N_6 = n_m(1+1)(1-0^2) = n_m \cdot 2 = 1$$

implying

$$n_m = \frac{1}{2}.$$

Hence, the mid-side shape functions are:

$$N_5 = \frac{1}{2}(1-u^2)(1-v),$$

$$N_6 = \frac{1}{2}(1+u)(1-v^2).$$

$$N_7 = \frac{1}{2}(1-u^2)(1+v).$$

$$N_8 = \frac{1}{2}(1-u)(1-v^2).$$

From here on, the process established in connection with the linear element is directly applicable. The nodal displacement vector will, of course, consist of 16 components and the N matrix of shape functions will also double in column size. The steps of the element matrix generation process are identical.

A similar flow of operations, in connection with the triangular element, results in a quadratic triangular element, the six-noded triangle. Let us consider the element depicted in Figure 1.4 and place mid-side nodes as follows:

$$(x_4, y_4) \rightarrow (1/2, 1/2),$$

$$(x_5, y_5) \rightarrow (0, 1/2),$$

and

$$(x_6, y_6) \rightarrow (1/2, 0).$$

Following above, the shape functions of the corner nodes will be

$$N_1 = (1-u-v)^2,$$

$$N_2 = u(2u-1),$$

and

$$N_3 = v(2v-1).$$

The mid-side nodes are represented by

$$N_4 = 4uv,$$

$$N_5 = 4(1 - u - v)v,$$

and

$$N_6 = 4u(1 - u - v).$$

They are of unit value in their respective locations and zero otherwise.

For higher order (so-called p-version) or physically more elaborate (non-planar) element formulations the reader is referred to [9] and [5], respectively.

The computational process of adding shape functions to mid-side nodes will easily generalize to three dimensions. The linear three dimensional elements, such as the linear tetrahedral element introduced in Section 3.4 and the linear hexahedral element, subject of Section 6.2 may be extended to quadratic elements by the same procedure.

The foundation established in this chapter should carry us into the modeling of a physical phenomenon, where one more generalization of the finite element technology will be done by addressing three-dimensional domains. Before this issue is explored, however, the generation of a finite element model is discussed.

References

[1] Clough, R. W.; The finite element method in plane stress analysis, Proceedings of 2nd Conference of electronic computations, ASCE, 1960

[2] Courant, P.; Variational methods for the solution of problems of equilibrium and vibration, Bulletin of American Mathematical Society, Vol. 49, pp. 1-23, 1943

[3] Galerkin, B. G.; Stäbe und Platten: Reihen in gewissen Gleichgewichtsproblemen elastischer Stäbe und Platten, Vestnik der Ingenieure, Vol. 19, pp. 897-908, 1915

[4] MacNeal, R. H.; NASTRAN theoretical manual, The MacNeal-Schwendler Corporation, 1972

[5] MacNeal, R. H.; Finite elements: Their design and performance, Marcel Dekker, New York, 1994

[6] Oden, J. T. and Reddy, J. N.; An introduction to the mathematical theory of finite elements, Wiley, New York, 1976

[7] Przemieniecki, J. S.; Theory of matrix structural analysis, McGraw-Hill, New York, 1968

[8] Ritz, W.; Über eine neue Methode zur Lösung gewisser Variationsprobleme der Mathematischen Physik, J. Reine Angewendte Mathematik, Vol. 135, pp. 1-61, 1908

[9] Szabo, B. and Babuska, I.; Finite element analysis, Wiley, New York, 1991

[10] Turner, M. J. et al; Stiffness and deflection analysis of complex structures, Journal of Aeronautical Science, Vol. 23, pp. 803-823, 1956

[11] Zienkiewicz, O. C.; The finite element method, McGraw-Hill, New York, 1968

2

Finite Element Model Generation

Finite element model generation involves two distinct components. First, the real life geometry of the physical phenomenon is approximated by geometric modeling. Second, the computational geometry model is discretized producing the finite element model. These issues are addressed in this chapter.

2.1 Bezier spline approximation

The first step in modeling the geometry of a solid object involves approximating its surfaces and edges with splines. Note, that this step also embodies a certain reduction as the real life continuum geometry is approximated by a finite number of computational geometry entities. The most popular and practical geometric modeling tools are based on parametric splines.

Let us first consider an edge of a physical model described by a curve whose equation is

$$\underline{r}(t) = x(t)\underline{i} + y(t)\underline{j} + z(t)\underline{k}.$$

The original curve will be approximated by a set of cubic parametric spline segments of form

$$S(t) = a + bt + ct^2 + dt^3,$$

where t ranges from 0.0 to 1.0. Let us assume a set of points $P_j, j = 1...m$, representing the geometric object we are to model. For simplicity let us focus on the first segment of the curve defined by four points P_0, P_1, P_2, P_3. These four points define a Bezier [1] polygon as shown in Figure 2.1. The curve will go through the end-points P_0 and P_3. The tangents of the curve at the end points will be defined by the two intermediate (control) points P_1, P_2.

The Bezier spline segment is formed from these four points as

$$S(t) = \Sigma_{i=0}^{3} P_i J_{3,i}(t).$$

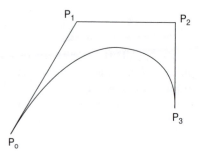

FIGURE 2.1 Bezier polygon

Here

$$J_{3,i}(t) = \binom{3}{i} t^i (1-t)^{3-i}$$

are binomial polynomials. Using the boundary conditions of the Bezier curve $(S(0), S(1), S'(0), S'(1))$ the matrix form of the Bezier spline segment may be written as

$$S(t) = TMP.$$

Here the matrix P contains the Bezier vertices

$$P = \begin{bmatrix} P_0 \\ P_1 \\ P_2 \\ P_3 \end{bmatrix},$$

and the matrix M the interpolation coefficients

$$M = \begin{bmatrix} 1 & 0 & 0 & 0 \\ -3 & 3 & 0 & 0 \\ 3 & -6 & 3 & 0 \\ -1 & 3 & -3 & 1 \end{bmatrix}.$$

T is a parametric row vector:

$$T = \begin{bmatrix} 1 & t & t^2 & t^3 \end{bmatrix}.$$

A very important generalization of this form is to introduce weight functions. The result is the rational parametric Bezier spline segment of form

$$S(t) = \frac{\Sigma_{i=0}^3 w_i P_i J_{3,i}(t)}{\Sigma_{i=0}^3 w_i J_{3,i}(t)},$$

or in matrix notation

$$S(t) = \frac{TM\overline{P}}{TMW}.$$

Here

$$\overline{P} = \begin{bmatrix} w_0 P_0 \\ w_1 P_1 \\ w_2 P_2 \\ w_3 P_3 \end{bmatrix}$$

is the vector of weighted point coordinates and

$$W = \begin{bmatrix} w_0 \\ w_1 \\ w_2 \\ w_3 \end{bmatrix}$$

is the array of weights. The weights have the effect of moving the curve closer to the control points, P_1, P_2, as shown in Figure 2.2.

The location of a specified point on the curve, P_s in Figure 2.2, defines three weights, while the remaining weight is covered by specifying the parameter value t^* to which the specified point should belong on the spline. Most commonly $t^* = \frac{1}{2}$ is chosen for such a point. The weights enable us to increase the fidelity of the approximation of the original curves.

The curve segment is finally approximated by

$$\underline{r}(t) = \frac{TM\overline{X}}{TMW}\underline{i} + \frac{TM\overline{Y}}{TMW}\underline{j} + \frac{TM\overline{Z}}{TMW}\underline{k}.$$

Here

$$\overline{X} = \begin{bmatrix} w_0 x_0 \\ w_1 x_1 \\ w_2 x_2 \\ w_3 x_3 \end{bmatrix}, \quad \overline{Y} = \begin{bmatrix} w_0 y_0 \\ w_1 y_1 \\ w_2 y_2 \\ w_3 y_3 \end{bmatrix}, \quad \overline{Z} = \begin{bmatrix} w_0 z_0 \\ w_1 z_1 \\ w_2 z_2 \\ w_3 z_3 \end{bmatrix},$$

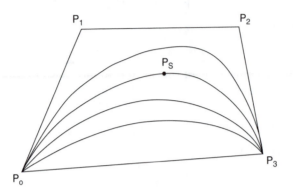

FIGURE 2.2 The effect of weights on the shape of spline

where x_i, y_i, z_i are the coordinates of the i-th Bezier point. An additional advantage of using rational Bezier splines is to be able to exactly represent conic sections and quadratic surfaces. These are common components of industrial models, for manufacturing as well as esthetic reasons.

In practice the geometric boundary is likely to be described by many points and therefore, a collection of spline segments. Consider the collection of points describing multiple spline segments shown in Figure 2.3. The most important question arising in this regard is the continuity between segments. Since the Bezier splines are always tangential to the first and last segments of the Bezier polygon, clearly a first order continuity exists only if the P_{i-1}, P_i, P_{i+1} points are collinear.

The presence of weights further specifies the continuity. Computing

$$\frac{\partial S}{\partial t}(t = 0) = 3\frac{w_1}{w_0}(P_1 - P_0)$$

and

$$\frac{\partial S}{\partial t}(t = 1) = 3\frac{w_2}{w_3}(P_3 - P_2).$$

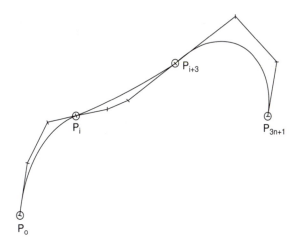

FIGURE 2.3 Multiple Bezier segments

Here $(P_i - P_j)$ is a vector pointing to P_j from P_i. Focusing on the adjoining segments of splines in Figure 2.4 the first order continuity condition is

$$\frac{w_{i-1}}{w_{i-0}}(P_i - P_{i-1}) = \frac{w_{i+1}}{w_{i+0}}(P_{i+1} - P_i).$$

There is a rather subtle but important distinction here. There is a geometric continuity component that means that the tangents of the neighboring spline segments are collinear. Then there is an algebraic component resulting in the fact that the magnitude of the tangent vectors is also the same. The notation w_{i+0}, w_{i-0} manifests the fact that the weights assigned to a control point in the neighboring segments do not have to be the same. If they are, a simplified first order continuity condition exists when

$$\frac{w_{i-1}}{w_{i+1}} = \frac{(P_{i+1} - P_i)}{(P_i - P_{i-1})}.$$

Enforcing such a continuity is important in the fidelity of the geometry approximation and in the discretization to be discussed later.

For the same reasons a second order continuity is also desirable. By definition

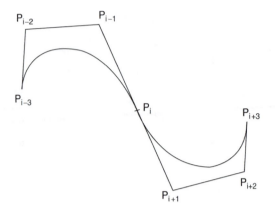

FIGURE 2.4 Continuity of spline segments

$$\frac{\partial^2 S}{\partial t^2}(t = 0) = (6\frac{w_1}{w_0} + 6\frac{w_2}{w_0} - 18\frac{w_1^2}{w_0^2})(P_1 - P_0) + 6\frac{w_2}{w_0}(P_2 - P_1)$$

and

$$\frac{\partial^2 S}{\partial t^2}(t = 1) = (6\frac{w_1}{w_3} + 6\frac{w_2}{w_3} - 18\frac{w_2^2}{w_3^2})(P_2 - P_3) + 6\frac{w_1}{w_3}(P_1 - P_2).$$

Generalization to the boundary of neighboring segments, assuming that the weights assigned to the common point between the segments is the same, yields the second order continuity condition as

$$w_{i-2}(P_{i-2} - P_i) - 3\frac{w_{i-1}^2}{w_i}(P_{i-1} - P_i) = w_{i+2}(P_{i+2} - P_i) - 3\frac{w_{i+1}^2}{w_i}(P_{i+1} - P_i).$$

This is a rather strict condition requiring that the two control points prior and after the common point (five points in all) are coplanar with some additional weight relations.

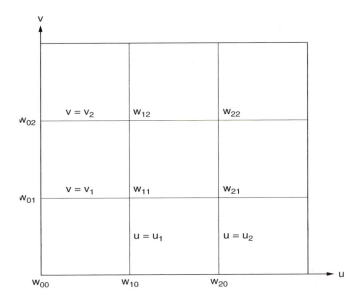

FIGURE 2.5 Bezier patch definition

2.2 Bezier surfaces

The method discussed above is easily generalized to surfaces. A Bezier surface patch is defined by a set of points on the surface of the physical model (plus the control points and weights) as shown in Figure 2.5. The rational parametric Bezier patch is described as

$$S(u,v) = \frac{\Sigma_{i=0}^{3}\Sigma_{j=0}^{3}w_{ij}J_{3,i}(u)J_{3,j}(v)P_{ij}}{\Sigma_{i=0}^{3}\Sigma_{j=0}^{3}w_{ij}J_{3,i}(u)J_{3,j}(v)}$$

or in matrix form

$$S(u,v) = \frac{UM\overline{P}M^{T}V}{UMWM^{T}V}.$$

The computational components are the matrix of weighted point coordinates

$$\overline{P} = \begin{bmatrix} w_{00}P_{00} & w_{01}P_{01} & w_{02}P_{02} & w_{03}P_{03} \\ w_{10}P_{10} & w_{11}P_{11} & w_{12}P_{12} & w_{13}P_{13} \\ w_{20}P_{20} & w_{21}P_{21} & w_{22}P_{22} & w_{23}P_{23} \\ w_{30}P_{30} & w_{31}P_{31} & w_{32}P_{32} & w_{33}P_{33} \end{bmatrix},$$

the parametric row vector of

$$U = \begin{bmatrix} 1 & u & u^2 & u^3 \end{bmatrix},$$

and column vector of

$$V = \begin{bmatrix} 1 \\ v \\ v^2 \\ v^3 \end{bmatrix}.$$

The weights form a matrix of:

$$W = \begin{bmatrix} w_{00} & w_{01} & w_{02} & w_{03} \\ w_{10} & w_{11} & w_{12} & w_{13} \\ w_{20} & w_{21} & w_{22} & w_{23} \\ w_{30} & w_{31} & w_{32} & w_{33} \end{bmatrix}.$$

The geometric surface of the physical model now is approximated by the patch of

$$\underline{r}(u,v) = \frac{UM\overline{X}M^T V}{TMWM^T V}\underline{i} + \frac{TM\overline{Y}M^T V}{TMWM^T V}\underline{j} + \frac{TM\overline{Z}M^T V}{TMWM^T V}\underline{k}.$$

Here $\overline{X}, \overline{Y}, \overline{Z}$, contain the weighted x, y, z point coordinates, respectively. Again, in a complex physical domain a multitude of these patches is used to completely cover the surface. The earlier continuity discussion generalizes for surface patches. The derivatives

$$\frac{\partial S(u,v)}{\partial u},$$

and

$$\frac{\partial S(u,v)}{\partial v}$$

will be the cornerstones of such relations. Similar arithmetic expressions used for the spline segments produce the first order continuity condition across the patch boundaries as shown in Figure 2.6.

$$\frac{w_{i+1,j+1}}{w_{i-1,j+1}} = \frac{(P_{i-1,j+1} - P_{i,j+1})}{(P_{i+1,j+1} - P_{i,j+1})}.$$

A similar treatment is applied to the v parametric direction. The second order

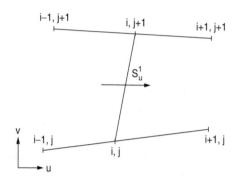

FIGURE 2.6 Patch continuity definition

continuity is based on

$$\frac{\partial^2 P(u, v)}{\partial u\, \partial v}$$

computed at the corners and the mathematics is rather tedious, albeit straightforward. The strictness of this condition is now almost overbearing, requiring nine control points to be the coplanar. Therefore, it is seldom enforced in geometric modeling for finite element applications. It mainly contributes to the esthetic appearance of the surface created and as such it is preferred by shape designers.

The technique also generalizes to volumes of the physical model as

$$S(u, v, t) = \frac{\Sigma_{i=0}^{3}\Sigma_{j=0}^{3}\Sigma_{k=0}^{3} w_{ijk} J_{3,i}(t) J_{3,j}(u) J_{3,k}(v) P_{ijk}}{\Sigma_{i=0}^{3}\Sigma_{j=0}^{3}\Sigma_{k=0}^{3} w_{ijk} J_{3,i}(t) J_{3,j}(u) J_{3,k}(v)}.$$

The matrix form and the final approximation form may be developed along the same lines as above for splines or patches. The result is

$$S(u, v, t) = \frac{\Sigma_{k=0}^{3} J_{3,i}(t) U M \overline{P}_k M^T V}{\Sigma_{k=0}^{3} J_{3,i}(t) U M W_k M^T V}.$$

The k layers of the volume are individual spline patches and the weights are defined as earlier. The formulation enables the modeling of volumes of revolutions or extrusions, details of those are beyond our needs here.

The points corresponding to equi-parametric values of the splines, surface patches and volumes are of course not equally separated in space. Sometimes it is necessary to re-parameterize one of these objects to smoothen the parametric distribution in a geometric sense. Nevertheless, the equi-parametric locations of these objects may constitute a basis for the discretization discussed in the next section.

The Bezier objects' industrial popularity is due to the following reasons:

1. The convex hull property: All Bezier curves, surface patches or volumes are contained inside of the hull of their control points,
2. The variation diminishing property: The number of intersection points between a Bezier curve and an infinite plane is the same as the number of intersections between the plane and the control polygon,
3. All derivatives and products of Bezier functions are easily computed Bezier functions.

These properties are exploited in industrial geometric modeling computations.

2.3   B-spline technology

An alternative to the Bezier spline technology is based on the B-splines. The technology allows a set of input points to be either interpolated or approximated, providing much more flexibility. The curves are still directed by control points, however, they are not given a priori, they are computed as part of the process. The technology, therefore, is more flexible than the Bezier technology and is preferred in the industry.

A general non-uniform, non-rational B-spline is described by

$$S(t) = \sum_{i=0}^{n} B_{i,k}(t)Q_i,$$

where Q_i are the yet unknown control points and $B_{i,k}$ are the B-spline basis functions of degree k. They are computed based on a certain parameterization influencing the shape of the curve. Note that for now we are focusing on non-rational, non-uniform B-splines.

The basis functions are initiated by

$$B_{i,0}(t) = \begin{cases} 1, t_i \leq t < t_{i+1} \\ 0, t < t_i, t \geq t_{i+1} \end{cases}$$

and higher order terms are recursively computed:

$$B_{i,k}(t) = \frac{t - t_i}{t_{i+k} - t_i} B_{i,k-1}(t) + \frac{t_{i+k+1} - t}{t_{i+k+1} - t_{i+1}} B_{i+1,k-1}(t).$$

The parameter values for the spline may be assigned via various methods. The simplest, and most widely used method is the uniform spacing. The method for $n + 1$ points is defined by the parameter vector

$$t = \begin{bmatrix} 0 \ 1 \ 2 \ \dots \ n \end{bmatrix}.$$

When the input points are geometrically somewhat equidistant this is proven to be a good method for parameterization. When the input points are spaced in widely varying intervals, a parameterization based on the chord length may also be used.

The parameter vector is commonly normalized as

$$t = \begin{bmatrix} 0 \ 1/n \ 2/n \ \dots \ 1 \end{bmatrix}.$$

Such normalization places all the parameter values in the interval $(0, 1)$ easing the complexity of the evaluation of the basis functions.

First we seek to **interpolate** a given set of points

$$P_j = (Px_j, Py_j, Pz_j); j = 0, \dots, m,$$

requiring that the B-spline ($S(t)$ at parameter value t_j passes through the given point P_j. This results in the equation

$$\begin{bmatrix} P_0 \\ P_1 \\ \dots \\ P_m \end{bmatrix} = \begin{bmatrix} B_{0,k}(t_0) & B_{1,k}(t_0) & B_{2,k}(t_0) & \dots & B_{n,k}(t_0) \\ B_{0,k}(t_1) & B_{1,k}(t_1) & B_{2,k}(t_1) & \dots & B_{n,k}(t_1) \\ \dots & \dots & \dots & \dots & \dots \\ B_{0,k}(t_m) & B_{1,k}(t_m) & B_{2,k}(t_m) & \dots & B_{n,k}(t_m) \end{bmatrix} \begin{bmatrix} Q_0 \\ Q_1 \\ \dots \\ Q_n \end{bmatrix}.$$

Using a matrix notation, the problem is

$$P = BQ,$$

where the P column matrix contains $m + 1$ terms and the Q column matrix contains $n+1$ terms, resulting in a rectangular system matrix B with $(m+1)$ rows and $(n + 1)$ columns. This problem may not be solved in general when

$m < n$, the case when the number of points given is less than the number of control points. The problem may also be only solved in a least squares sense when $m > n$, having more input points than control points.

The problem has a unique solution for the case of $m = n$ and in this case the sequence of unknown control points is obtained in the form of

$$Q = B^{-1}P,$$

where the inverse is shown for the sake of simplicity, it is not necessarily computed. In fact, the B matrix exhibits a banded pattern that is dependent on the degree k of the spline chosen. Specifically, the semi-bandwidth is less than the order k.

$$B_{i,k}(t_j) = 0; \, for \, |i - j| >= k.$$

This fact should be exploited to produce an efficient solution.

The second approach is to **approximate** the input points in a least squares sense, resulting in a distinctly different curve. This may be obtained by finding a minimum of the squares of the distances between the spline and the points.

$$\sum_{j=0}^{m} (S(t_j) - P_j)^2.$$

Substituting the B-spline formulation and the basis functions results in

$$\sum_{j=0}^{m} (\sum_{i=0}^{n} B_{i,k}(t_j)Q_i - P_j)^2.$$

The derivative with respect to an unknown control point Q_p is

$$2\sum_{j=0}^{m} B_{p,k}(t_j)(\sum_{i=0}^{n} B_{i,k}(t_j)Q_i - P_j) = 0,$$

where $p = 0, 1, \ldots, n$. This results in a system of equations, with $n + 1$ rows and columns, in the form:

$$B^T BQ = B^T P$$

with the earlier introduced B matrix. The solution of this system produces an approximated, not interpolated solution.

The technology may also be extended to include smoothing considerations and directional constraints to the splines, topics that are discussed at length in [3].

2.4 Computational example

Considering that the problem is given in 3-space, the solution for the x, y, z coordinates may be obtained simultaneously.

$$
\begin{bmatrix}
Qx_0 & Qy_0 & Qz_0 \\
Qx_1 & Qy_1 & Qz_1 \\
\cdots \\
Qx_n & Qy_n & Qz_n
\end{bmatrix}
= B^{-1}
\begin{bmatrix}
Px_0 & Py_0 & Pz_0 \\
Px_1 & Py_1 & Pz_1 \\
\cdots \\
Px_n & Py_n & Pz_n
\end{bmatrix}.
$$

For a fixed degree, say $k = 3$, and uniformly parameterized B-spline segments the basis functions may be analytically computed as:

$$
B_{0,3} = \frac{1}{6}(1 - t)^3,
$$

$$
B_{1,3} = \frac{1}{6}(3t^3 - 6t^2 + 4),
$$

$$
B_{2,3} = \frac{1}{6}(-3t^3 + 3t^2 + 3t + 1),
$$

and

$$
B_{3,3} = \frac{1}{6}t^3.
$$

For the case of 4 points $(n = 3)$ the uniform parameter vector becomes:

$$
t = \begin{bmatrix} 0 & 1 & 2 & 3 \end{bmatrix}.
$$

For this case the interpolation system matrix is easily computed by hand as

$$
B = \frac{1}{6}
\begin{bmatrix}
1 & 4 & 1 & 0 \\
0 & 1 & 4 & 1 \\
-1 & 4 & -5 & 8 \\
-8 & 31 & -44 & 27
\end{bmatrix}.
$$

The matrix is positive definite and its inverse is:

$$
B^{-1} = \frac{1}{6}
\begin{bmatrix}
21 & -28 & 17 & -4 \\
4 & 5 & -4 & 1 \\
-1 & 8 & -1 & 0 \\
0 & -1 & 8 & -1
\end{bmatrix}.
$$

The solution for the control points is obtained as

$$
Q = B^{-1}P,
$$

where P is the vector of input points. For example for the points

$$P = \begin{bmatrix} 0 & 0 \\ 1 & 1 \\ 2 & 1 \\ 3 & 0 \end{bmatrix},$$

the control points obtained are

$$Q = \begin{bmatrix} -1 & -11/6 \\ 0 & 1/6 \\ 1 & 7/6 \\ 2 & 7/6 \end{bmatrix}.$$

Figure 2.7 shows the curve generated from the control points **interpolating** the given input points, while spanning the parameter range from 0 to 3.

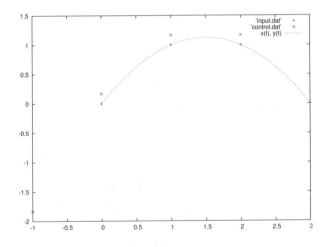

FIGURE 2.7 B spline interpolation

To evaluate the spline curve as function of any parameter value in the span, the following matrix formula (conceptually similar to the Bezier form) may be used:

$$S(t) = TCQ,$$

with

$$C = \frac{1}{6} \begin{bmatrix} 1 & 4 & 1 & 0 \\ -3 & 0 & 3 & 0 \\ 3 & -6 & 3 & 0 \\ -1 & 3 & -3 & 1 \end{bmatrix}$$

where the C matrix is gathered from the coefficients of the analytic basis functions above, and

$$T = \begin{bmatrix} 1 & t & t^2 & t^3 \end{bmatrix}.$$

This formula enables the validation of the spline going through the input points. For example

$$S_{t=1} = TCQ = \begin{bmatrix} 1 & 1 \end{bmatrix},$$

which of course agrees with the second input point.

For demonstration of the **approximation** computation, we add another input point to the above set. The given set of 5 input points are:

$$P = \begin{bmatrix} 0 & 0 \\ 1 & 1 \\ 2 & 1 \\ 3 & 0 \\ 2 & -1 \end{bmatrix}.$$

For the case of 5 points ($n = 4$) the parameter vector becomes:

$$t = \begin{bmatrix} 0 & 1 & 2 & 3 & 4 \end{bmatrix}.$$

For this case the B matrix is

$$B = \frac{1}{6} \begin{bmatrix} 1 & 4 & 1 & 0 \\ 0 & 1 & 4 & 1 \\ -1 & 4 & -5 & 8 \\ -8 & 31 & -44 & 27 \\ -27 & 100 & -131 & 64 \end{bmatrix}.$$

The solution for the control points in this case is obtained as

$$Q = (B^T B)^{-1} B^T P.$$

Figure 2.8 shows the curve generated from these control points **approximating** the given input points, while spanning the parameter range from 0 to 4. The evaluation yields the approximation points

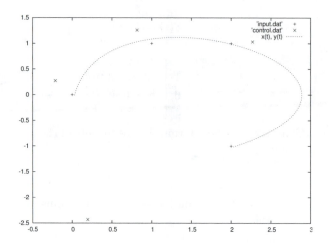

FIGURE 2.8 B spline approximation

$$
S_{app} = \begin{bmatrix}
0.028571 & -0.014286 \\
0.885714 & 1.057143 \\
2.171429 & 0.914286 \\
2.885714 & 0.057143 \\
2.028571 & -1.014286
\end{bmatrix},
$$

which reasonably well approximate the input points, while producing a smooth curve.

The selection of the parameter values enables interesting and useful shape variations of the spline around the same set of given points. For example, repeated values of the parameter values at both ends enforce a clamped boundary condition, forcing the curve through the end points. Figure 2.9 shows the curve generated from these control points and for the same input points, still approximating them.

Note that the number of pre-assigned parameter values, in this case becomes 9 and the parameter vector will be

$$
t = \begin{bmatrix} 0\ 0\ 0\ 1/4\ 1/2\ 3/4\ 1\ 1\ 1 \end{bmatrix}.
$$

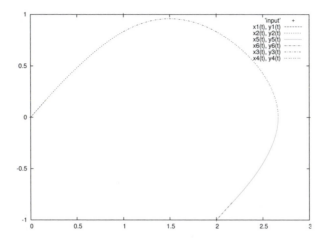

FIGURE 2.9 Clamped B spline approximation

Also note that the number of sections of the spline increased to 6 in this case. The figure depicts the sections of the spline with different line patterns as shown on the legend, while spanning the parameter range from 0 to 1.

Finally, the curve adhering to the same set of input points may also be closed by repeating an input point. The starting point repeated at the end results in the closed curve shown in Figure 2.10, in this case with 5 sections.

This example also demonstrated that a level continuity between the segments of the B-spline is automatically assured depending on the degree k of the spline. There is no need for special considerations when large number of input points are given.

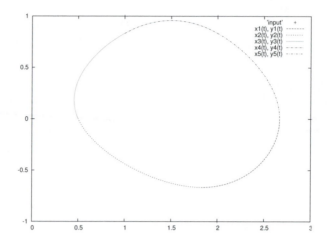

FIGURE 2.10 Closed B spline approximation

2.5   NURBS objects

As in the Bezier technology, it is also possible to use weights in the B-spline technology, resulting in rational B-splines. When a non-uniform parameterization is also used, the splines become Non-Uniform, Rational B-splines, known as NURBS.

Introducing weights associated with each control point results in the NURBS curve of form

$$S(t) = \frac{\sum_{i=0}^{n} w_i B_{i,k}(t) Q_i}{\sum_{i=0}^{n} w_i B_{i,k}(t)}.$$

The geometric meaning of the weights is similar to that of the Bezier technology, they will pull the curve closer to the input points. It is, however, important to point out that changing one single weight value will result only in a local shape change in the segment related to the point. This local control is one of the advantages of the B-spline technology over the Bezier approach.

The formulation extends quite easily to surfaces:

$$S(u,v) = \frac{\sum_{i=0}^{n}\sum_{j=0}^{m} w_{i,j} B_{i,k}(u) B_{j,l}(v) Q_{i,j}}{\sum_{i=0}^{n}\sum_{j=0}^{m} w_{i,j} B_{i,k}(u) B_{j,l}(v)}.$$

Note that the degree of the v directional parametric curve may be different than that of the u curve, denoted by l. Similarly the parameterization in both directions may be different. This gives tremendous flexibility to the method.

Geometric modeling operations are enabled by these objects. Consider generating a swept surface by moving a curve $C(u)$ along a trajectory $T(v)$. This is conceptually similar to generating a cylinder by defining a circle and the axis perpendicular to the plane of the circle. In general, the surface generated by this process may be described as

$$S(u,v) = C(u) + T(v).$$

Assume that the curves are NURBS of the same order

$$C(u) = \frac{\sum_{i=0}^{n} w_i^C B_{i,k}(u) Q_i^C}{\sum_{i=0}^{n} w_i^C B_{i,k}(u)}$$

and

$$T(v) = \frac{\sum_{j=0}^{m} w_j^T B_{j,k}(v) Q_j^T}{\sum_{j=0}^{m} w_j^T B_{j,k}(v)}.$$

Then the swept NURBS surface is of form

$$S(u,v) = \frac{\sum_{i=0}^{n}\sum_{j=0}^{m} w_{i,j} B_{i,k}(u) B_{j,l}(v) Q_{i,j}}{\sum_{i=0}^{n}\sum_{j=0}^{m} w_{i,j} B_{i,k}(u) B_{j,k}(v)},$$

where

$$Q_{i,j} = Q_i^C + Q_j^T$$

and

$$w_{i,j} = w_i^C w_j^T.$$

Similar considerations may be used to generate NURBS surfaces of revolution around a given axis.

Finally, the NURBS also generalize to three dimensions for modeling volumes:

$$S(u,v,t) = \frac{\sum_{i=0}^{n}\sum_{j=0}^{m}\sum_{p=0}^{q} w_{i,j,p} B_{i,k}(u) B_{j,k}(v) B_{p,k}(t) Q_{i,j,p}}{\sum_{i=0}^{n}\sum_{j=0}^{m}\sum_{p=0}^{q} w_{i,j,p} B_{i,k}(u) B_{j,k}(v) B_{p,k}(t)}.$$

The form is written with the assumption of the curve degree being the same (k) in all three parametric directions, albeit that is not necessary.

Finally, it is important to point out that the surface representations via either Bezier or B-splines may produce non-rectangular surface patches. Such, for example triangular, patches are very important in the finite element discretization step to be discussed next. They may easily be produced from above formulations by collapsing a pair of points into one and will not be discussed further [3].

2.6   Geometric model discretization

The foundation of many general methods of discretization (commonly called meshing) is the classical Delaunay triangulation method [4]. The Delau-

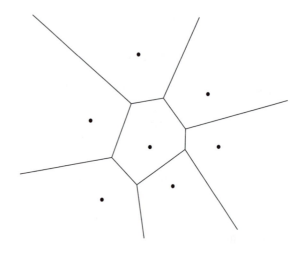

FIGURE 2.11 Voronoi polygon

nay triangulation technique in turn is based on Voronoi polygons [8] . The Voronoi polygon, assigned to a certain point of a set of points in the plane, contains all the points that are closer to the selected point than to any other point of the set.

Let us define the set of points $S \subseteq R^2$ and $P_i \in S$ be the points of the set $i = 1, 2, ..n$. The points $Q(x, y) \in R^2$ that satisfy

$$\|Q(x, y) - P_i\| \leq \|Q(x, y) - P_j\|, \forall P_j \in S,$$

constitute the Voronoi polygon $V(P_i)$ of point P_i. The Voronoi polygon is a convex polygon.

The inequalities represent half planes between point P_i and every point P_j. The intersection of these half planes produces the Voronoi polygon. For example consider the set of points shown in Figure 2.11. The irregular hexagon containing one point in the middle (the P_i point) is the Voronoi polygon of point P_i.

It is easy to see that the points inside the polygon $(Q(x, y))$ are closer to P_i than to any other points of the set. It is also quite intuitive that the edges of the Voronoi polygon are the perpendicular bisectors of the line segments connecting the points of the set.

The union of the Voronoi polygons of all the points in the set completely covers the plane. It follows that the Voronoi polygon of two points of the set do not have common interior points; at most they share points on their common boundary.

The definition and process generalizes to three dimensions very easily. If the set of points are in space, $S \subseteq R^3$, the points $Q(x, y, z) \in R^3$ that satisfy

$$\|Q(x, y, z) - P_i\| \leq \|Q(x, y, z) - P_j\|, \forall P_j \in S,$$

define the Voronoi polyhedron $V(P_i)$ of P_i.

Every inequality defines a half-space and the Voronoi polyhedron $V(P_i)$ is the intersection of all the half-spaces defined by the point set. The Voronoi polyhedron is a convex polyhedron.

2.7 Delaunay mesh generation

The Delaunay triangulation process is based on the Voronoi polygons as follows. Let us construct Delaunay edges by connecting points P_i and P_j when their Voronoi polygons $V(P_i)$ and $V(P_j)$ have a common edge. Constructing all such possible edges will result in the covering of the planar region of our interest with triangular regions, the Delaunay triangles.

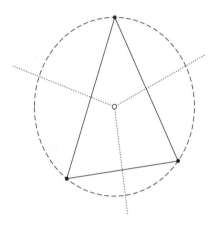

FIGURE 2.12 Delaunay triangle

Figure 2.12 shows a Delaunay triangle. The dotted lines are the edges of the Voronoi polygons and the solid lines depict the Delaunay edges. The process extends quite naturally and covers the plane as shown in Figure 2.13 with 6 Delaunay triangles. It is known that under the given definitions no two Delaunay edges cross each other.

On the other hand it is possible to have a special case when four (or even more) Voronoi polygons meet at a common point. This degenerate case will result in the Delaunay edges producing a quadrilateral. As the discretized regions are the finite elements for our further computations, this case is no cause for panic. We can certainly have quadrilateral finite elements as was shown earlier. There are also remedies to preserve a purely triangular mesh; slightly moving one of the points participating in the scenario will eliminate

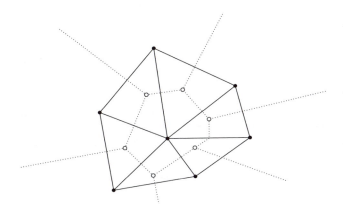

FIGURE 2.13 Delaunay triangularization

the special case.

Finally, in three dimensions the Delaunay edges are defined as lines connecting points that share a common Voronoi facet (a face of a Voronoi polyhedron). Furthermore, the Delaunay facets are defined by points that share a common Voronoi edge (an edge of a Voronoi polyhedron). In general each edge is shared by exactly three Voronoi polyhedron, hence the Delaunay regions' facets are going to be triangles.

The Delaunay regions connect points of Voronoi polyhedra that share a common vertex. Since in general the number of such polyhedra is four, the generated Delaunay regions will be tetrahedra. The Delaunay method generalized into three dimensions is called Delaunay tessellation [6].

There are many automatic methods to discretize a two-dimensional, not necessarily planar, domain. [2] describes such a method for surface meshing with rectangular elements. There are also other methods in the industry to partition a three-dimensional domain into a collection of non-overlapping elements that covers the entire solution domain, see for example [7]. The most successful techniques are the proprietary heuristic algorithms used in commercial software. The quality of the mesh heavily influences the finite element solution results. A good quality mesh has elements close to equal in size with shapes that are not too distorted. In the case of hexahedron elements this means element shapes that approach cubes. Gross inequality in the ratios of the sides (called aspect ratio in the industry) results in less accurate solutions.

The final topic of the finite element model generation is the assignment of node numbers. This step will influence the topology of the assembled finite element matrices, and as such, it influences the computational performance. The finite element matrix reordering is discussed in Section 7.1.

The assignment of the node numbers usually starts at a corner or an edge of the geometric model, now meshed, and proceeds inward towards the interior of the model while at the same time considering the element connectivity. The goal of this is that nodes of an element should have neighboring numbers. It is not necessary to achieve that, but is it useful as a pre-processing for reordering and assuring that operation's efficiency.

2.8   Model generation case study

To demonstrate the model generation process we consider a simple engineering component of a bracket. This example will be used in the last section also as a case study for a complete engineering analysis. The process in today's engineering practice is almost exclusively executed in a computer aided design (CAD) software environment. The advantage of working in such environment is that the engineer is able to immediately analyze the model, since the model is created in a computer.

Still, the engineer starts by creating a design sketch, such as shown in Figure 2.14. The role of the design sketch is to specify the contours of the desired shape that will accommodate the kinematic relationships between the component and the rest of the product. Since this is an interactive process, the engineer could easily modify the sketch until it satisfies the goals.

The next step in the design process is to "fill out" the details of the geometry. The model may be extruded from two dimensional contour elements in a certain direction, or blended between contour curves. The interior volumes may be filled with standard geometrical components like cylinders or cones. The process usually entails generating the faces and interior volumes of the model from many components. Figure 2.15 depicts the geometric model of

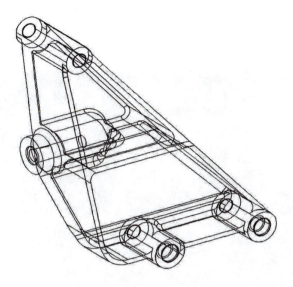

FIGURE 2.14 Design sketch of a bracket

the bracket example.

The geometric modeling software environment facilitates the easy execution of coordinate transformations, such as rotations, translations of reflections, enabling the engineer to try various scenarios. Earlier designs may be reused and modifications easily made to produce a variant product. Since shape is really independent of frame of reference, this approach encapsulates the shape in a parametric form, not in a fixed reference frame of the blue-prints of the past. Another advantage of the parametric representation is the easy re-sizing of the model.

Finally the finite element discretization step is executed on the geometric model using the techniques described in the last two sections. This step is nowadays fully automated and produces mostly triangular surface and tetra-hedral volume meshes, such as visible in Figure 2.16.

For complex models consisting of multitudes of surface and volume components, the various sections may be meshed separately and the boundary regions are re-meshed to achieve mesh-continuity. This approach is also advantageous from a computational performance point of view, since the separate sections may be meshed simultaneously on multiprocessor computers.

In most cases the finite element sizes are strongly influenced by the small-

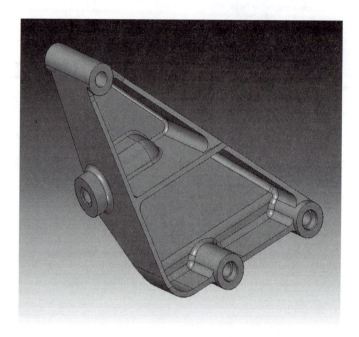

FIGURE 2.15 Geometric model of bracket

est geometric features of the geometric model and this may result in denser mesh in other areas of the geometry. For instance, it is noticeable in the example that the fillet surface between the cylinder on the left and the facing planar side of the model seem to have dictated the mesh size. This approach may be detrimental to the solution performance when there are structurally unimportant minor details, such as esthetic components of the structure. The modeling and meshing of such do not necessarily improve the quality of the results either.

The physical problem is still not yet fully defined by the geometric and the finite element models. The modeling of the physical phenomenon, such as elasticity, and the specification of the material properties of the part need to be executed. These topics are the subject of the next chapter.

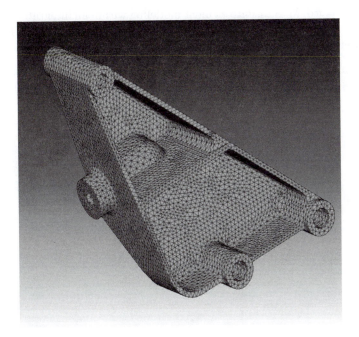

FIGURE 2.16 Finite element model of bracket

References

[1] Bezier, P.; Essai de definition numerique des courbes et de surfaces experimentals, Universite D. et. M. Curie, Paris, 1977

[2] Blacker, T. D. and Stephenson, M. B.; Paving: A new approach to automated quadrilateral mesh generation, Report DE-AC04-17DP00789, Sandia National Laboratory, 1990

[3] Gregory, J. A.; The mathematics of surfaces, Springer, New York, 1978

[4] Delaunay, B.; Sur la sphere vide, Izv. Akad. Nauk SSSR, Otdelenie Matematicheskih i Estestvennyh Nauk, Vol. 7, pp. 793-800, 1934

[5] Komzsik, L.; Applied variational analysis for engineers, Taylor and Francis, Boca Raton, 2009

[6] Shenton, D. N. and Cendes, Z. J.; Three-dimensional finite element mesh generation using Delaunay tessellation, IEEE Trans. Magn., Vol. 21, pp. 2535-2538, 1985

[7] Shephard, M. S.; Finite element modeling within an integrated geometric modeling environment: Part I - Mesh generation, Report TR-85024, Renssealer Polytechnic Institute, 1985

[8] Voronoi, G.; Nouvelles applications des paramêtres continus â la thêorie des formes quadratiques. J. Reine Angew. Math., Vol 133, pp. 97-178, 1907

3

Modeling of Physical Phenomena

A mechanical system will be used to present the modeling of a physical problem with the finite element technique. The techniques presented in the book are, however, applicable in many other engineering principles as shown in [2].

3.1 Lagrange's equations of motion

The analysis of a mechanical system is based on Lagrange's equations of motion of analytic mechanics, see [1] and [3] for various formulations. The equations are

$$\frac{d}{dt}\frac{\partial T}{\partial \dot{q}_i} - \frac{\partial T}{\partial q_i} + \frac{\partial P}{\partial q_i} + \frac{\partial D}{\partial \dot{q}_i} = 0, \; i = 1, ..n.$$

The q_i are generalized coordinates, describing the motion of the mechanical system. Here T is the kinetic energy and P is the potential energy, while D is the dissipative function of the system.

The potential energy consists of the internal strain energy (E_s) and work potential (W_p) of external forces as

$$P = E_s + W_p.$$

For demonstration consider the simple discrete mass-spring system shown in Figure 3.1. The (only) generalized coordinate describing the motion of the system is the only degree of freedom, the displacement in the x direction, $q_1 = x$. The kinetic energy of the system is related to the motion of the mass as

$$T = \frac{1}{2}m\dot{x}^2.$$

The strain energy here is the energy stored in the spring and it is

$$E_s = \frac{1}{2}kx^2.$$

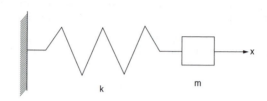

FIGURE 3.1 Discrete mechanical system

The total potential energy is

$$P = E_s.$$

Appropriate differentiation yields

$$\frac{d}{dt}\frac{\partial T}{\partial \dot{x}} = m\ddot{x}$$

and

$$\frac{\partial P}{\partial x} = kx.$$

Substituting into Lagrange's equation of motion produces the well-known equation of

$$m\ddot{x} + kx = 0.$$

This equation, the archetype example used in the study of ordinary differential equations, describes the undamped free vibrations of a single degree of freedom mass-spring discrete mechanical system. For the damped and forced vibration cases see [6] for example.

3.2 Continuum mechanical systems

A continuum mechanical system with a general geometry is usually analyzed in terms of the displacements of its particles. The displacements of the particles of the continuum are $q = q(x, y, z)$ where x, y, z are the geometric coordinates of the particle in space. The finite element discretization of the three-dimensional continuum model leads to a set of nodes. They are like the nodes in the two-dimensional example in the prior chapter, however, with an added spatial dimension.

The kinetic energy of a continuum system in terms of the particle velocities is

$$T = \frac{1}{2} \int \dot{q}^T \dot{q} \rho dV,$$

where ρ is the mass per unit volume. The internal strain energy for the continuum mechanical system is

$$E_s = \frac{1}{2} \int \sigma^T \epsilon dV.$$

Here σ, ϵ are the stresses and strains of the system. The work potential is

$$W_p = - \int q^T f_A dV,$$

where f_A contains the active forces acting on the system. Finally the dissipative function is

$$D = \frac{1}{2} \int \dot{q}^T f_D \dot{q} dV,$$

where f_D contains dissipative forces. The active forces are associated with the q displacements and the dissipative forces with the $\dot{q}$ velocities. The dissipative forces usually, and the active forces sometimes, act on the surface of the mechanical systems, however, we assume here for simplicity that they have been converted to a volume integral.

Let us consider a mechanical system with node points having six degrees of freedom. We assume that they are free to move in all three spatial directions and rotate freely around all three axes. This is true in standard analysis of structures and the infrastructure of some commercial finite element codes is

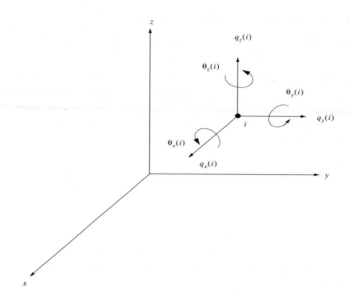

FIGURE 3.2 Degrees of freedom of mechanical particle

built around such assumption. For the i-th node point

$$v(i) = \begin{bmatrix} q_x(i) \\ q_y(i) \\ q_z(i) \\ \theta_x(i) \\ \theta_y(i) \\ \theta_z(i) \end{bmatrix},$$

where $q_x(i), q_y(i), q_z(i)$ are the translational degrees of freedom of the ith node point and $\theta_x(i), \theta_y(i), \theta_z(i)$ are the rotational degrees of freedom, as shown in Figure 3.2.

The complete mechanical model is described with a vector v, that contains all the node displacements of all node points as

$$v = \begin{bmatrix} v(1) \\ v(2) \\ . \\ v(i) \\ . \\ v(n) \end{bmatrix},$$

where n is the number of discrete node points, hence the order of v is $g = 6n$ where g is the total number of degrees of freedom in the model.

3.3 Finite element analysis of elastic continuum

The purpose of this book is to discuss the computational techniques of finite element analysis that are applicable to various physical principles. Therefore, while the concepts will have a mechanical foundation, they will also carry over to other principles where a potential field is the basis of describing the physical phenomenon.

For example, in heat conduction the potential function is the temperature field (t), the "strain" is the temperature gradient $(-\nabla t)$ and the "stress" is the heat flow (q). Similarly in magneto-statics the magnetic vector potential (A) is the fundamental "displacement" field, and the physical equivalent to the strain is the magnetic induction (B). The "stress" in magneto-statics is the magnetic field strength (H).

Similar analogies exist with other physical disciplines. It is important to point out that the most important quantity from the engineers perspective varies from discipline to discipline. For the heat conduction engineer the temperature field (the "displacement") is of primary interest and it is luckily the primary result of the finite element solutions. For the electrical engineer studying magneto-statics the magnetic induction (the "strain") is the most important. Finally, for the structural engineer both stresses and displacements are of practical interest.

These major concepts are now defined in connection with an elastic continuum. The physical behavior of an elastic continuum is analyzed via the the relative displacements on neighboring points in the body. The relative displacements represent the physical strain inside the body. The strains were already discussed in Chapter 1 as a mathematical concept, here we give a physical foundation.

There are two distinct types of physical strains:

a. Extensional strains, and

b. Shear strains.

The first kind of strain is the change in distance between two points of the body. The shear strain is defined as the change in the angle between two lines which were perpendicular in the undeformed body. The strain vector of these distinct components for the general 3-dimensional model is

$$\epsilon = \begin{bmatrix} \frac{\partial q}{\partial x} \\ \frac{\partial q}{\partial y} \\ \frac{\partial q}{\partial z} \\ \frac{\partial q}{\partial x} + \frac{\partial q}{\partial y} \\ \frac{\partial q}{\partial y} + \frac{\partial q}{\partial z} \\ \frac{\partial q}{\partial z} + \frac{\partial q}{\partial x} \end{bmatrix}.$$

The physical stress components of the body corresponding to these strains are:

$$\sigma = \begin{bmatrix} \sigma_x \\ \sigma_y \\ \sigma_z \\ \tau_{xy} \\ \tau_{yz} \\ \tau_{zx} \end{bmatrix}.$$

Here the σ are the normal and the τ are the shear stresses. The stress-strain relationship is described by

$$\sigma = D\epsilon,$$

where the D matrix describes the constitutive relationship due to the elastic properties of the model, such as Poisson's ratio and Young's modulus of elasticity. The actual structure of D depends on the material modeling. For a general three-dimensional model of linear elastic, isotropic material the D matrix is of form

$$D = \frac{E}{(1+\nu)(1-2\nu)} \begin{bmatrix} 1-\nu & \nu & \nu & 0 & 0 & 0 \\ \nu & 1-\nu & \nu & 0 & 0 & 0 \\ \nu & \nu & 1-\nu & 0 & 0 & 0 \\ 0 & 0 & 0 & 0.5-\nu & 0 & 0 \\ 0 & 0 & 0 & 0 & 0.5-\nu & 0 \\ 0 & 0 & 0 & 0 & 0 & 0.5-\nu \end{bmatrix}.$$

Here E is the Young's modulus and ν is the Poisson ratio.

It was established earlier that the strains are related to the node displacements as

$$\epsilon = Bq_e,$$

therefore, the stresses are also related via

$$\sigma = DBq_e.$$

The structure of the B matrix for general three-dimensional continuum problems will be discussed in detail shortly. For a three-dimensional element, the element strain energy is formed as

$$E_e = \frac{1}{2} \int \int \int \sigma^T \epsilon dx dy dz = \frac{1}{2} q_e^T k_e q_e^T,$$

where the element stiffness matrix is

$$k_e = \int \int \int B^T D B dx dy dz.$$

To find more details on the continuum mechanics foundation [5] is the classical reference.

3.4 A tetrahedral finite element

To demonstrate the finite element modeling of a three-dimensional continuum we consider the tetrahedron element, such as shown in Figure 3.3 located generally in global coordinates. This is the most common object resulting from the discretization process discussed in Sections 2.6 and 2.7, albeit not the most advantageous from numerical perspective. Every node point of the tetrahedron has three degrees of freedom, they are free to move in the three spatial directions. Hence, there are twelve nodal displacements of the element as

$$q_e = \begin{bmatrix} q_{1x} \\ q_{1y} \\ q_{1z} \\ q_{2x} \\ q_{2y} \\ q_{2z} \\ q_{3x} \\ q_{3y} \\ q_{3z} \\ q_{4x} \\ q_{4y} \\ q_{4z} \end{bmatrix}.$$

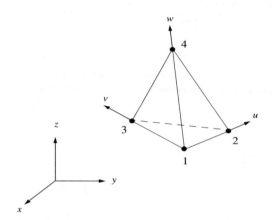

FIGURE 3.3 Tetrahedron element

Note, that in element derivation sessions throughout, we use the notation of q_{ix} to refer to the x translation of the ith local node of the element. This is a distinction from the notation of $q_x(i)$ which refers to the x translation of the ith global node of the model.

   The displacement at any location inside this element is approximated with the help of the shape functions as

$$q(x, y, z) = N q_e.$$

Generalizing the observation of the relationship between the shape functions and the parametric coordinates with respect to the triangular element, we may use

$$N_1 = w, \; N_2 = u, \; N_3 = v,$$

and

$$N_4 = 1 - u - v - w.$$

Here the parametric coordinate w is from the 1st node to the 4th node. This arbitrary selection is chosen to be consistent with the triangular element introduced in Chapter 1. Then u is from the 1st node to the 2nd node and finally, the parametric coordinate v is from the 1st node to the 3d node. Such a selection of the N_i functions obviously satisfies

$$N_1 + N_2 + N_3 + N_4 = 1.$$

We organize the N matrix of the four shape functions as

$$N = \begin{bmatrix} N_1 & 0 & 0 & N_2 & 0 & 0 & N_3 & 0 & 0 & N_4 & 0 & 0 \\ 0 & N_1 & 0 & 0 & N_2 & 0 & 0 & N_3 & 0 & 0 & N_4 & 0 \\ 0 & 0 & N_1 & 0 & 0 & N_2 & 0 & 0 & N_3 & 0 & 0 & N_4 \end{bmatrix}.$$

Then

$$\begin{bmatrix} q_x(x,y,z) \\ q_y(x,y,z) \\ q_z(x,y,z) \end{bmatrix} = \begin{bmatrix} N_1 & 0 & 0 & N_2 & 0 & 0 & N_3 & 0 & 0 & N_4 & 0 & 0 \\ 0 & N_1 & 0 & 0 & N_2 & 0 & 0 & N_3 & 0 & 0 & N_4 & 0 \\ 0 & 0 & N_1 & 0 & 0 & N_2 & 0 & 0 & N_3 & 0 & 0 & N_4 \end{bmatrix} \begin{bmatrix} q_{1x} \\ q_{1y} \\ q_{1z} \\ q_{2x} \\ q_{2y} \\ q_{2z} \\ q_{3x} \\ q_{3y} \\ q_{3z} \\ q_{4x} \\ q_{4y} \\ q_{4z} \end{bmatrix},$$

or

$$q(x,y,z) = N q_e$$

as desired. The location of a point inside the element is approximated again with the same four shape functions as the displacement field

$$x = N_1 x_1 + N_2 x_2 + N_3 x_3 + N_4 x_4,$$

$$y = N_1 y_1 + N_2 y_2 + N_3 y_3 + N_4 y_4,$$

and

$$z = N_1 z_1 + N_2 z_2 + N_3 z_3 + N_4 z_4.$$

Here x_i, y_i, z_i are the x, y, z coordinates of the i-th node of the tetrahedron, hence the element is again an iso-parametric element. Using the above definition of the shape functions with the element coordinates and by substituting we get

$$x = x_4 + (x_1 - x_4)w + (x_2 - x_4)u + (x_3 - x_4)v,$$

$$y = y_4 + (y_1 - y_4)w + (y_2 - y_4)u + (y_3 - y_4)v,$$

and

$$z = z_4 + (z_1 - z_4)w + (z_2 - z_4)u + (z_3 - z_4)v.$$

These equations will be fundamental to the integral transformation of the element matrix generation for the tetrahedron element. Assuming that the B, D matrices are constant in an element (constant strain element), the element stiffness matrix is formulated as

$$k_e = B^T DB \int \int \int dxdydz = B^T DBV_e,$$

where V_e is the volume of the element as

$$V_e = \int \int \int dxdydz.$$

The integral will be again transformed to the parametric coordinates

$$V_e = \int \int \int det[\frac{\partial(x,y,z)}{\partial(u,v,w)}]dudvdw.$$

Here the Jacobian matrix is

$$\frac{\partial(x,y,z)}{\partial(u,v,w)} = \begin{bmatrix} \frac{\partial x}{\partial u} & \frac{\partial x}{\partial v} & \frac{\partial x}{\partial w} \\ \frac{\partial y}{\partial u} & \frac{\partial y}{\partial v} & \frac{\partial y}{\partial w} \\ \frac{\partial z}{\partial u} & \frac{\partial z}{\partial v} & \frac{\partial z}{\partial w} \end{bmatrix} = \begin{bmatrix} x_2 - x_4 & x_3 - x_4 & x_1 - x_4 \\ y_2 - y_4 & y_3 - y_4 & y_1 - y_4 \\ z_2 - z_4 & z_3 - z_4 & z_1 - z_4 \end{bmatrix} = J,$$

and it is visibly constant for the element.

Gaussian integration requires the additional transformation of this integral, as integrating over the parametric domain of our element, the integral has the following boundaries

$$\int_{u=0}^{1} \int_{v=0}^{1-u} \int_{w=0}^{1-u-v} f(u,v,w)dwdvdu = \int_{r=-1}^{1} \int_{s=-1}^{1} \int_{t=-1}^{1} f(s,r,t)dtdrds.$$

The Gaussian integration method applied to this triple integral is

$$\int_{r=-1}^{1} \int_{s=-1}^{1} \int_{t=-1}^{1} f(s,r,t)dtdrds = \Sigma_{i=1}^{n} c_i \Sigma_{j=1}^{n} c_j \Sigma_{k=1}^{n} c_k f(s_i, r_j, t_k).$$

With applying the one or two point formula, the element volume is

$$V_e = \frac{1}{6}det(J).$$

To complete the element matrix, the strain components and, in turn, terms of the B matrix still need to be computed. Clearly

$$\begin{bmatrix} \frac{\partial q}{\partial u} \\ \frac{\partial q}{\partial v} \\ \frac{\partial q}{\partial w} \end{bmatrix} = J \begin{bmatrix} \frac{\partial q}{\partial x} \\ \frac{\partial q}{\partial y} \\ \frac{\partial q}{\partial z} \end{bmatrix}.$$

Then

$$\begin{bmatrix} \frac{\partial q}{\partial x} \\ \frac{\partial q}{\partial y} \\ \frac{\partial q}{\partial z} \end{bmatrix} = J^{-1} \begin{bmatrix} \frac{\partial q}{\partial u} \\ \frac{\partial q}{\partial v} \\ \frac{\partial q}{\partial w} \end{bmatrix}$$

The terms of J^{-1} may be computed as

$$J^{-1} = \frac{adj(J)}{det(J)} = \begin{bmatrix} b_{11} & b_{12} & b_{13} \\ b_{21} & b_{22} & b_{23} \\ b_{31} & b_{32} & b_{33} \end{bmatrix}.$$

With the introduction of terms

$$b_1 = -(b_{11} + b_{12} + b_{13}),$$

$$b_2 = -(b_{21} + b_{22} + b_{23}),$$

and

$$b_3 = -(b_{31} + b_{32} + b_{33}),$$

the terms of the B matrix are

$$B = \begin{bmatrix} b_{11} & 0 & 0 & b_{12} & 0 & 0 & b_{13} & 0 & 0 & b_1 & 0 & 0 \\ 0 & b_{21} & 0 & 0 & b_{22} & 0 & 0 & b_{23} & 0 & 0 & b_b & 0 \\ 0 & 0 & b_{31} & 0 & 0 & b_{32} & 0 & 0 & b_{33} & 0 & 0 & b_3 \\ 0 & b_{31} & b_{32} & 0 & b_{32} & b_{22} & 0 & b_{33} & b_{23} & 0 & b_3 & b_2 \\ b_{31} & 0 & b_{11} & b_{32} & 0 & b_{12} & b_{33} & 0 & b_{13} & b_3 & 0 & b_1 \\ b_{21} & b_{11} & 0 & b_{22} & b_{12} & 0 & b_{23} & b_{13} & 0 & b_2 & b_1 & 0 \end{bmatrix}.$$

$B^T DB$ produces a 12×12 element matrix. This element matrix will contribute to 12 columns and rows of the assembled finite element matrix.

3.5 Equation of motion of mechanical system

The motion of the particles of the continuum mechanical system is approximated by the discrete node displacements as

$$q = Nv,$$

where N is a collection of shape functions shown earlier. The kinetic energy with this approximation becomes

$$T = \frac{1}{2} \dot{v}^T \Sigma_{e=1}^m \int N^T N \rho dV_e \dot{v},$$

where m is the number of elements and e is the element index. The integral in the above equation is performed on each element. The shape of the element and the connectivity of the related nodes is represented in the N shape functions. Executing the summation for all the finite elements of the model we get

$$T = \frac{1}{2}\dot{v}^T M \dot{v},$$

where M is the mass matrix. It is computed similarly to the stiffness matrix as

$$M = \Sigma_{e=1}^m m_e$$

and

$$m_e = \int N^T N \rho dV_e,$$

using the same numerical integration and local-global transformation principles as discussed earlier. Similar manipulations on the potential energy yield

$$P = \frac{1}{2}v^T K v - v^T F,$$

where K is the stiffness matrix. It is computed according to the procedure developed earlier. The F is the vector of all active forces (volume and surface) and computed as

$$F = \Sigma_{e=1}^m f_e,$$

where f_e is the element force. The differentiation of the kinetic energy yields

$$\frac{\partial T}{\partial \dot{v}} = M \dot{v}$$

and

$$\frac{d}{dt}(M\dot{v}) = M\ddot{v}.$$

Note, that T only depends of $\dot{v}$, so the second term of Lagrange's equations of motion (the derivative of T with respect to v) is ignored. Similarly the potential energy terms yield:

$$\frac{\partial P}{\partial v} = Kv - F.$$

Here the K stiffness matrix generation was detailed in the last section. The dissipative function becomes

$$D = \frac{1}{2}\dot{v}^T B \dot{v},$$

and differentiation yields

$$\frac{\partial D}{\partial \dot{v}} = B\dot{v}.$$

Here B is the damping matrix and not the strain-displacement matrix. Substituting all the above forms into Lagrange's equations of motion we obtain the matrix equation of the equilibrium of a general mechanical system

$$M\ddot{v}(t) + B\dot{v}(t) + Kv(t) = F(t),$$

where v is the displacement vector of the system at time t. In Chapter 14 we will address the direct solution of this problem in the time domain. This is a second order, non-homogeneous, ordinary differential equation with matrix coefficients. The generally time-dependent active loads are contained in the right-hand side matrix.

For our discussion we will mostly assume that the coefficient matrices are constant; hence the equation is linear. In practice the coefficient matrices are often not constant, material and geometric nonlinearities exist, see for example, [4] for more details. The computational components of such cases will be discussed in Chapter 16, however, the computational techniques developed until then are also applicable to them.

We will also assume in most of the following that the matrices are real and symmetric. This restriction will be released in Chapter 10, where the complex spectral computation will also be addressed.

The computational techniques discussed in the bulk of the book with these restrictions in mind are just as applicable.

3.6 Transformation to frequency domain

In many cases the equilibrium equation is transformed from the time domain to the frequency domain. This is accomplished by a Fourier transformation of form

$$u(\omega) = \int_0^\infty v(t)e^{-i\omega t}\,dt.$$

Assuming zero initial conditions, the result is the algebraic equation of motion

$$(-\omega^2 M + i\omega B + K)u(\omega) = F(\omega).$$

This equation describes the most general problem of the forced vibrations of a damped structure. It is called the frequency response analysis equation in some commercial software, and the harmonic analysis in some other programs. It has become the most widely used type of analysis because it is more

repeatable and less costly than transient response analysis. The application
of the Fourier transformation to the right-hand side results in

$$F(\omega) = \int_0^\infty F(t)e^{-i\omega t}dt.$$

Once the frequency domain solution is obtained, the time domain response
may be computed as

$$v(t) = \frac{1}{\pi}\int_0^\infty Re(u(\omega)e^{i\omega t})d\omega.$$

Depending on the absence of certain matrices in the algebraic equation of motion, various other structural engineering analysis problems are formulated.

The simplest case is when there is no mass (M) or damping (B) in the mechanical model and the load, hence the solution is not frequency-dependent.
This is the case of linear static analysis which computes the static equilibrium
of the system. The algebraic equation is simply

$$Ku = F.$$

The solution vector u describes the spatial coordinates of each discrete variable due to the static load F. The computational components of the solution
of the problem are presented in Chapter 7. The reduction technique applicable to this class of problems is the static condensation of Chapter 8.

Another distinct class of analysis is the free vibration of structures. In this
class, there are no external loads acting on the structure, i.e., F does not
exist. The class is further subdivided into damped and undamped cases. The
algebraic equation of the free vibration of an undamped system is

$$(K - \lambda M)u(\omega) = 0, \quad \lambda = \omega^2.$$

This very important problem, called normal modes analysis in the industry,
will be addressed in Chapter 9. The dynamic reduction technique of Chapter
11 and the modal synthesis technique discussed in Chapter 12 are also used
for the solution of this problem.

A very specific subcase of this is the buckling analysis. The goal of the
buckling analysis is to find the critical load under which a structure may become unstable. The problem is formulated as

$$(K - \lambda_b K_g)u(\lambda_b) = 0,$$

where the K_g matrix is the so-called geometric or differential stiffness matrix.
The geometric stiffness matrix depends on the geometry of the structure and
the applied load. The buckling eigenvalue λ_b is a scaling factor. Multiplying

the applied load (that was used to compute the geometric stiffness) by λ_b will produce a critical load for the structure. The corresponding vector of $u(\lambda_b)$ is the buckling shape.

The stiffness and the mass matrix terms are generally derived with mathematical rigor. Damping terms, by contrast, are developed more for engineering convenience. One type, called structural damping is defined by multiplying the terms of the stiffness matrix by ig, where i is the imaginary unit and g is a parameter based on engineering experience, with 0.03 being a typical value. The g values may be applied to the element stiffness matrix terms, or the assembled stiffness matrix terms, or both.

The structure may also contain actual damper components, such as hydraulic shock absorbers that are always modeled in the damping matrix. The damping level is one of the main parameters in determining the amount of amplification of response at resonance. Considering the above, free vibrations of damped systems are the solutions of

$$((1 + ig)K + i\omega B - \omega^2 M)u(\omega) = 0.$$

For types of analyses where imaginary stiffness terms are inconvenient, such as transient response analysis, they may be converted into equivalent terms of the damping matrix B. This conversion is somewhat dubious, especially since the structural damping may not capture all the damping phenomena, such as the play in riveted or bolted joints, flexing of welds, or the thickness of adhesives in bonded structures. Nevertheless, in the following the damping will always be contained in the B matrix.

The most practical class of analysis, however, is the forced vibration of structures. In this case external loads act on the structure, i.e. $F \neq 0$. This class is also subdivided into damped and undamped cases. The algebraic equation of the forced vibration of an undamped system is

$$(K - \omega^2 M)u(\omega) = F(\omega),$$

where ω is an excitation frequency. The forced vibration of a damped system is described by

$$(K + i\omega B - \omega^2 M)u(\omega) = F(\omega).$$

These problems of frequency response analysis, will be addressed in Chapter 15.

These equations, however, are far from ready to be solved. The steps of modifying the equations of motion usually reduce the size of the matrices. In order to follow the change of size in the reduction process, in the following, the matrices will always have their sizes indicated in double subscripts. The vectors whose size is not indicated by a single subscript are assumed to be compatible with the corresponding matrices. For example, the normal modes equation in this form is

$$(K_{gg} - \lambda M_{gg})u_g(\omega) = 0,$$

indicating that the matrices are the yet-unreduced global matrices with $g = 6 * n$ columns and rows, and the solution vector with the same number of rows. Here n is the number of node points.

In order to obtain a good numerical solution from the equations, the issue of processing constraints as well as various boundary conditions must be addressed. These are usually applied by the engineer directly or by the choice of the applied modeling technique and discussed next.

References

[1] Béda, Gy. and Bezák, A.; Kinematika és dinamika, Tankönyvkiadó, Budapest, 1969

[2] Brauer, J. R.; What every engineer should know about finite element analysis, Marcel Dekker, 1993

[3] Lanczos, C.; The variational principles of mechanics, Dover, Mineola, 1979

[4] Oden, J. T.; Finite elements of nonlinear continua, McGraw-Hill, New York, 1972

[5] Timoshenko, S. and Goodier, J. N.; Theory of elasticity, McGraw-Hill, New York, 1951

[6] Vierck, R. K.; Vibration analysis, International Textbook Co., Scranton, 1967

4

Constraints and Boundary Conditions

Constraint equations may specify the coefficients for flexible elements or for a completely assembled finite element model. A constraint equation states that the displacements at a selected set of degrees of freedom must satisfy an equation of constraint independently of the force-deflection relationships defined by the flexible elements. There may be many constraint equations and each may enforce different laws.

Commercial finite element codes often have methods for defining the coefficients of the constraint equation with element-like input formats because the coefficients are tedious to develop by hand, and can cause implausible results when a decimal point is in the wrong location, or a sign is changed. These types of errors in constraint equation coefficients are difficult to identify by scanning input files.

One class of an element-like constraint equation generator is for rigid elements whose degrees of freedom move as if they were attached to a rigid component. Another type is an interpolation element, where a set of degrees of freedom is attached to a reference node in a manner that the motion of the reference node is a weighted combination of the motion of the other connected degrees of freedom.

Some programs also allow input of constraint equation coefficients directly as well, to allow analysts to implement features that are not built into the program. For example component mode synthesis (the topic of Chapter 12) may also be achieved by appropriately defined constraint equations.

The subject of this chapter is the mathematics of describing and applying (removing or augmenting) these constraints, most commonly named multipoint constraints. The bibliography contains some publications related to this topic, see [1], [3] and [4].

4.1   The concept of multi-point constraints

Multiple degrees of freedom constraints are usually applied before the single degree of freedom constraints in commercial applications. The reason for this is that the application of the multiple degrees of freedom constraints removes some of the singularities of the global matrix. The issue of dealing with the single degree of freedom constraints representing boundary conditions is then simpler as it is related only to the remaining singularities in the model.

Let us consider a very simple mechanical problem of a bar in the $x - y$ plane connecting two node points as shown in Figure 4.1. The constraints of the bar will be discussed in this chapter and the solution of the problem will culminate in Chapter 5.

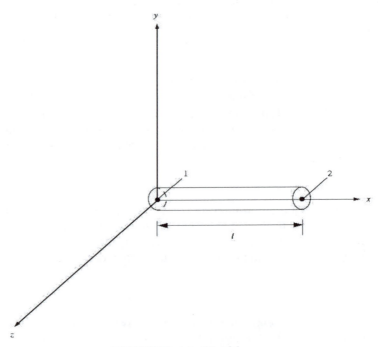

FIGURE 4.1 Rigid bar

Let us assume the bar is rigid under bending but flexible under axial elongation, such an element is sometimes called an axial bar. This restriction will be released in Chapter 17 with the introduction of a planar bending of the bar.

The bar is of length l and its local x axis is aligned with its axis. As this element constitutes the whole example model, the global degree of freedom notation of $q_x(i)$ will be used.

The motion of nodes 1 and 2 is obviously related. The y displacement of node 2 depends on the motion of node 1 as

$$q_y(2) = q_y(1) + l\theta_z(1),$$

where $\theta_z(1)$ is the rotation of node 1 with axis of rotation being the z coordinate axis. It is assumed to be a small rotation such that the

$$sin(\theta_z(1)) \approx \theta_z(1)$$

assumption is valid. Similarly the z displacement is

$$q_z(2) = q_z(1) + l\theta_y(1).$$

Let us consider the case of $l = 1$ for simplicity, without limiting the generality of the discussion. The rotations with respect to the out of plane z axis (resulting from the rigidity) are obviously identical

$$\theta_z(2) = \theta_z(1).$$

Similarly the rotations with respect to the y axis are related as

$$\theta_y(2) = \theta_y(1).$$

Finally, assuming that the bar is rigid with respect to torsion also, we have another constraint of

$$\theta_x(2) = \theta_x(1).$$

Note, that these are already multi-point constraints as "point" in this context means the degrees of freedom, not the nodes. One can write the constraint equations in the form of

$$q_y(1) + \theta_z(1) - q_y(2) = 0,$$
$$q_z(1) + \theta_y(1) - q_z(2) = 0,$$
$$\theta_x(1) - \theta_x(2) = 0,$$
$$\theta_y(1) - \theta_y(2) = 0,$$

and

$$\theta_z(1) - \theta_z(2) = 0.$$

Note, that the length variable is now ignored as it is assumed to be unity, and the equations are reordered according to the dependent degrees of freedom. A more general form of writing these equations in connection with the global finite element model is

$$R_{mg} u_g = 0,$$

where R_{mg} is the constraint matrix and m is the number of constraints. In the simple bar example that is a five by twelve matrix as we have five constraints:

$$R_{mg} = \begin{bmatrix} 0 & 1 & 0 & 0 & 0 & 1 & 0 & -1 & 0 & 0 & 0 & 0 \\ 0 & 0 & 1 & 0 & 1 & 0 & 0 & 0 & -1 & 0 & 0 & 0 \\ 0 & 0 & 0 & 1 & 0 & 0 & 0 & 0 & 0 & -1 & 0 & 0 \\ 0 & 0 & 0 & 0 & 1 & 0 & 0 & 0 & 0 & 0 & -1 & 0 \\ 0 & 0 & 0 & 0 & 0 & 1 & 0 & 0 & 0 & 0 & 0 & -1 \end{bmatrix}.$$

The structure of u_g is simply an order twelve column vector as we have two nodes each consisting of six degrees of freedom:

$$u_g = \begin{bmatrix} q_x(1) \\ q_y(1) \\ q_z(1) \\ \theta_x(1) \\ \theta_y(1) \\ \theta_z(1) \\ q_x(2) \\ q_y(2) \\ q_z(2) \\ \theta_x(2) \\ \theta_y(2) \\ \theta_z(2) \end{bmatrix}.$$

We can further partition the solution vector as

$$u_g = \begin{bmatrix} u_n \\ u_m \end{bmatrix},$$

where the m partition contains the degrees of freedom whose motion depends on the motion of the n partition degrees of freedom. Hence, the m and n degrees of freedom are called dependent and independent degrees of freedom, respectively. More on this in the next section. In the simple bar example it is also easy to identify these partitions as the dependent motions of node 2:

$$u_m = \begin{bmatrix} q_y(2) \\ q_z(2) \\ \theta_x(2) \\ \theta_y(2) \\ \theta_z(2) \end{bmatrix},$$

and the independent motions of nodes 1 and 2:

$$u_n = \begin{bmatrix} q_x(1) \\ q_y(1) \\ q_z(1) \\ \theta_x(1) \\ \theta_y(1) \\ \theta_z(1) \\ q_x(2) \end{bmatrix}.$$

Although trivial in the bar example, this partitioning is representative of multi-point constraints.

In practical applications more difficult rigid elements, such as levers, mechanisms and gear trains, are frequently applied. These constraints connect several nodes and there may be many of them. Therefore, a practical finite element analysis model may be subject to many multi-point constraints resulting in a large number of independent constraint equations. Note, that the number of equations is not necessarily the same as the number of physical constraints as some of these may produce more than one equation.

The remainder of this chapter focuses on the reduction step eliminating these constraints.

4.2 The elimination of multi-point constraints

In a general form, the problem of eliminating the multi-point constraints is equivalent to solving the quadratic minimization problem

$$min\Pi = u_g^T K_{gg} u_g - u_g^T F_g,$$

subject to the equality constraints

$$R_{mg} u_g = 0$$

where Π is the functional to be minimized (in stationary problems the potential energy). The R_{mg} matrix is the coefficient matrix of the linear equality constraints, representing relationships amongst the variables required to be enforced. The reduction involves the selection of m linearly independent vectors from the columns of R_{mg}.

Let us partition the R_{mg} matrix accordingly into the m and n partitions:

$$R_{mg} = \left[\, R_{mn} \; R_{mm} \,\right].$$

Note, that the *a priori* identification of the m (linearly independent) and the n partitions is somewhat arbitrary and in general cases requires some engineering intuition. The dependent degrees of freedom (m partition) define a matrix made from a subset of the constraint equations that must be well conditioned for matrix inversion.

The reduction technique related to the multi-point constraints is as follows. The m partition will be removed from further computations. The solution will be obtained in the remaining n partition. Finally the m partition's displacements are computed from the n partition solution. Hence, the m partition degrees of freedom "depend" on the n partition degrees of freedom, as was mentioned in the prior section. Therefore, they are called dependent degrees of freedom in the industry, although they actually form a linearly independent set. By the same token the n partition degrees of freedom are called the independent set. In the following this industrial notation is used.

Finding the linearly independent m partition of the R_{mg} matrix is a standard linear algebra problem. In part it may be solved by purely mathematical considerations. For example, a degree of freedom may not be dependent in more than one constraint equation. It is also important to identify redundant constraint equations.

There can be a lot of computational work required to pick the dependent set, particularly when there are many over-lapping constraint equations. Most of the commercial finite element tools provide efficient techniques to automatically select the dependent set.

In industrial practice, there are some other considerations. For example in case of structural analysis, the degrees of freedom of substructures that are to be coupled with other substructures may not be eliminated because the elimination makes them unavailable for use as boundary points. These, application specific decisions may also be automated, although the underlying rules must be made by the engineer.

In the following discussion we assume that the partitioning is properly chosen. The multi-point constraint reduction will be facilitated by the matrix

$$G_{mn} = -\left[R_{mm}^{-1} \right] \left[R_{mn} \right].$$

Let us consider the linear statics problem partitioned accordingly

$$\begin{bmatrix} \overline{K}_{nn} & K_{nm} \\ K_{mn} & K_{mm} \end{bmatrix} \begin{bmatrix} u_n \\ u_m \end{bmatrix} = \begin{bmatrix} \overline{F}_n \\ F_m \end{bmatrix},$$

where the bar over certain partitions is for notational convenience. The original K_{gg} matrix is assumed to be symmetric, meaning that $K_{nm} = K_{mn}^T$. Since

$$\begin{bmatrix} R_{mn} & R_{mm} \end{bmatrix} \begin{bmatrix} u_n \\ u_m \end{bmatrix} = 0,$$

using the definition of G_{mn} it follows that

$$u_m = G_{mn} u_n.$$

Introducing this to the last equation yields

$$\begin{bmatrix} \overline{K}_{nn} & K_{nm} G_{mn} \\ K_{mn} & K_{mm} G_{mn} \end{bmatrix} \begin{bmatrix} u_n \\ u_n \end{bmatrix} = \begin{bmatrix} \overline{F}_n \\ F_m \end{bmatrix}.$$

Pre-multiplying the second equation by G_{mn}^T to maintain symmetry we get

$$\begin{bmatrix} \overline{K}_{nn} & K_{nm} G_{mn} \\ G_{mn}^T K_{mn} & G_{mn}^T K_{mm} G_{mn} \end{bmatrix} \begin{bmatrix} u_n \\ u_n \end{bmatrix} = \begin{bmatrix} \overline{F}_n \\ G_{mn}^T F_m \end{bmatrix}.$$

The simultaneous solution of the two equations by summing them yields

$$K_{nn} u_n = F_n.$$

The reduced n-size stiffness matrix is built as

$$K_{nn} = \overline{K}_{nn} + K_{nm} G_{mn} + G_{mn}^T K_{mn} + G_{mn}^T K_{mm} G_{mn}.$$

It is important to note that since we are ultimately solving equilibrium equations, the matrix modifications need to be reflected on the right-hand sides also:

$$F_n = \overline{F}_n + G_{mn}^T F_m.$$

In practical circumstances, depending on the industry, the g partition to n partition reduction may be as much as 20-30 percent. The larger numbers are characteristic of the automobile industry due to very specific techniques such as modeling spot welds with multi-point constraints.

For the simple bar example, the calculation is as follows.

$$R_{mm} = \begin{bmatrix} -1 & 0 & 0 & 0 & 0 \\ 0 & -1 & 0 & 0 & 0 \\ 0 & 0 & -1 & 0 & 0 \\ 0 & 0 & 0 & -1 & 0 \\ 0 & 0 & 0 & 0 & -1 \end{bmatrix}.$$

$$R_{mn} = \begin{bmatrix} 0 & 1 & 0 & 0 & 0 & 1 & 0 \\ 0 & 0 & 1 & 0 & 1 & 0 & 0 \\ 0 & 0 & 0 & 1 & 0 & 0 & 0 \\ 0 & 0 & 0 & 0 & 1 & 0 & 0 \\ 0 & 0 & 0 & 0 & 0 & 1 & 0 \end{bmatrix}.$$

Since $R_{mm} = -I$ is non-singular and equal to its inverse,

$$G_{mn} = - \begin{bmatrix} 0 & 1 & 0 & 0 & 0 & 1 & 0 \\ 0 & 0 & 1 & 0 & 1 & 0 & 0 \\ 0 & 0 & 0 & 1 & 0 & 0 & 0 \\ 0 & 0 & 0 & 0 & 1 & 0 & 0 \\ 0 & 0 & 0 & 0 & 0 & 1 & 0 \end{bmatrix}.$$

4.3   An axial bar element

Before we proceed further, let us derive the stiffness matrix for the simple bar element shown in Figure 4.1, allowing for axial deformation. The local element coordinate system is as earlier, aligned with the axis of the element. The following derivation is aimed at the specific example we used for the constraint elimination process.

Using earlier principles we introduce a local coordinate u with the following convention: $u = 0$ at node 1 and $u = 1$ at node 2. We describe the coordinates of the points of the element with

$$x = N_1 x_1 + N_2 x_2 = (1 - u)x_1 + ux_2.$$

Here x_i are the node point coordinates.

Using the iso-parametric concept, the same shape functions are used to describe the deformation of the element as

$$q(x) = N_1 q_{1x} + N_2 q_{2x} = (1 - u)q_{1x} + uq_{2x}.$$

With

$$N = \begin{bmatrix} 1 - u & u \end{bmatrix},$$

and

$$q_e = \begin{bmatrix} q_{1x} \\ q_{2x} \end{bmatrix},$$

we have

$$q(x) = N q_e.$$

Observe the local displacement notation.

The strain in the element is

$$\epsilon = \frac{dq}{dx} = \frac{dq}{du}\frac{du}{dx}.$$

Differentiation yields

$$\frac{dq}{du} = q_{2x} - q_{1x}$$

and

$$\frac{dx}{du} = x_{2x} - x_{1x} = l,$$

where l is the length of the element. Hence, the strain is

$$\epsilon = \frac{q_2 - q_1}{l}.$$

The strain-displacement relationship of

$$\epsilon = B q_e$$

is satisfied with the strain-displacement matrix

$$B = \frac{1}{l}\begin{bmatrix} -1 & 1 \end{bmatrix}.$$

Note again, that when using linear shape functions the strain will be constant within an element.

For this one-dimensional tension-compression problem the elastic behavior is defined by the well-known Hooke's law

$$\sigma = E\epsilon,$$

where E is the Young's modulus of elasticity. The stiffness matrix for the element is

$$k_e = \int \int \int B^T E B dx dy dz.$$

Considering that the bar has a constant cross section A:

$$k_e = AE \int_{x_1}^{x_2} B^T B dx.$$

Introducing the local coordinates and substituting yields

$$k_e = \frac{AE}{l^2} \int_0^1 \begin{bmatrix} -1 \\ 1 \end{bmatrix} \begin{bmatrix} -1 & 1 \end{bmatrix} J du.$$

With the Jacobian

$$J = \frac{dx}{du} = l$$

the element stiffness matrix finally is

$$k_e = \begin{bmatrix} a & -a \\ -a & a \end{bmatrix},$$

where

$$a = \frac{AE}{l}.$$

This element stiffness matrix will contribute to the 1st and 7th column of the assembled matrix, resulting in

$$K_{gg} = \begin{bmatrix} a & 0 & 0 & 0 & 0 & 0 & -a & 0 & 0 & 0 & 0 & 0 \\ 0 & 0 & 0 & 0 & 0 & 0 & 0 & 0 & 0 & 0 & 0 & 0 \\ 0 & 0 & 0 & 0 & 0 & 0 & 0 & 0 & 0 & 0 & 0 & 0 \\ 0 & 0 & 0 & 0 & 0 & 0 & 0 & 0 & 0 & 0 & 0 & 0 \\ 0 & 0 & 0 & 0 & 0 & 0 & 0 & 0 & 0 & 0 & 0 & 0 \\ 0 & 0 & 0 & 0 & 0 & 0 & 0 & 0 & 0 & 0 & 0 & 0 \\ -a & 0 & 0 & 0 & 0 & 0 & a & 0 & 0 & 0 & 0 & 0 \\ 0 & 0 & 0 & 0 & 0 & 0 & 0 & 0 & 0 & 0 & 0 & 0 \\ 0 & 0 & 0 & 0 & 0 & 0 & 0 & 0 & 0 & 0 & 0 & 0 \\ 0 & 0 & 0 & 0 & 0 & 0 & 0 & 0 & 0 & 0 & 0 & 0 \\ 0 & 0 & 0 & 0 & 0 & 0 & 0 & 0 & 0 & 0 & 0 & 0 \\ 0 & 0 & 0 & 0 & 0 & 0 & 0 & 0 & 0 & 0 & 0 & 0 \end{bmatrix}.$$

The other degrees of freedom are undefined. Applying the elimination process to the stiffness matrix results in the following reduced, 7 by 7 matrix:

$$K_{nn} = \begin{bmatrix} a & 0 & 0 & 0 & 0 & 0 & -a \\ 0 & 0 & 0 & 0 & 0 & 0 & 0 \\ 0 & 0 & 0 & 0 & 0 & 0 & 0 \\ 0 & 0 & 0 & 0 & 0 & 0 & 0 \\ 0 & 0 & 0 & 0 & 0 & 0 & 0 \\ 0 & 0 & 0 & 0 & 0 & 0 & 0 \\ -a & 0 & 0 & 0 & 0 & 0 & a \end{bmatrix}.$$

After this elimination step the recovery of the engineering solution set is formally obtained by

$$u_g = \begin{bmatrix} I_{nn} \\ G_{mn} \end{bmatrix} u_n,$$

where the actual merging of the very likely interleaved partitions is not visible. This will be clarified when executing this step for our example in the next chapter.

Let us summarize the multi-point constraint elimination step as follows:

$$\begin{bmatrix} K_{gg} & \\ & [\, K_{nn} \,] \end{bmatrix}.$$

The special notation in this chart (and similar charts at the end of later chapters) represents the fact that the original K_{gg} matrix has been reduced to the K_{nn} matrix. It is not a mathematical expression in the sense, that K_{nn} is not a direct partition of K_{gg}.

The topic of the next two sections is the application of boundary conditions to the finite element model. These are usually enforced displacements of the finite element model. These are represented by single degree of freedom constraints, commonly called single-point constraints. These constraints are in essence a sub-case of the more general method of the multi-point constraints presented in the last sections. Their separate presentation in these sections is justified because they are also handled separately in the industry and they are related to the automatic detection of singularities presented in the next chapter.

4.4   The concept of single-point constraints

The single-point constraints apply a fixed value to a degree of freedom of a certain node, hence the name. The most common occurrence of such is, the description of boundary conditions, i.e., the fixed (often zero) displacement of a node in a certain direction or its rotation by a certain axis.

Let us consider our simple example again, but constrain the translation of node 1, imposing a boundary condition via the following constraint equations

$$q_x(1) = 0.0,$$

$$q_y(1) = 0.0,$$

and

$$q_z(1) = 0.0.$$

These are clearly single-point constraints as they constrain one single degree of freedom each. Their physical meaning is to keep node 1 fixed at the origin, represented by the shading at the left of the bar in Figure 4.2.

Some commercial software programs automatically assign six degrees of freedom per node point. Some types of models have fewer degrees of freedom per node points. Most solid elements have only translation degrees of freedom leaving only three degrees of freedom per node point. The stiffness matrix columns for the other degrees of freedom are null, meaning that the structure is not defined in these degrees of freedom. Single-point constraints can be used to remove these undefined degrees of freedom. Other possible uses for single point constraints, such as to enforce symmetry of the deformation of a symmetric structure or component, are not discussed here.

4.5   The elimination of single-point constraints

The n-size equilibrium equation may further be reduced by applying the single point constraints. The single-point constraints are described by

$$u_s = Y_s,$$

where the u_s is a partition of u_n:

$$u_n = \begin{bmatrix} u_f \\ u_s \end{bmatrix}.$$

For our simple example

$$u_s = \begin{bmatrix} q_x(1) \\ q_y(1) \\ q_z(1) \end{bmatrix},$$

leaving the unconstrained set

$$u_f = \begin{bmatrix} \theta_x(1) \\ \theta_y(1) \\ \theta_z(1) \\ q_x(2) \end{bmatrix}.$$

The Y_s vector of enforced generalized displacements (translations and rotations) for our example is:

$$Y_s = \begin{bmatrix} 0.0 \\ 0.0 \\ 0.0 \end{bmatrix}.$$

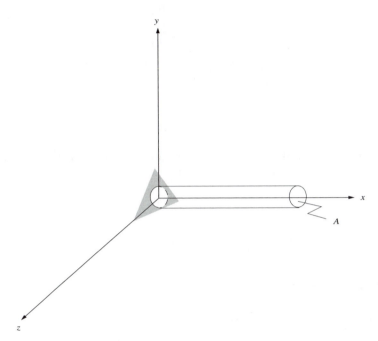

FIGURE 4.2 Boundary conditions

Note, that the components are not always zero; sometimes a non-zero displacement or rotation value is enforced by the engineer. This so-called enforced motion case will be discussed in Section 15.5. The corresponding reduction of the stiffness matrix is based on the appropriate partitioning of the last reduced equation:

$$K_{nn}u_n = \begin{bmatrix} K_{ff} & K_{fs} \\ K_{sf} & K_{ss} \end{bmatrix} \begin{bmatrix} u_f \\ u_s \end{bmatrix} = \begin{bmatrix} F_f \\ F_s \end{bmatrix}.$$

As the second equation has a prescribed solution, the remaining equation is

$$K_{ff}u_f = F_f - K_{fs}Y_s.$$

The f partition is sometimes called the free partition, as the constraints have been eliminated. The size reduction between the n and f partitions is not as dramatic as between the g and the n partitions. Usually there are only a few boundary conditions even for very large problems.

This step for our example results in the further reduced, now 4 by 4 matrix:

$$K_{ff} = \begin{bmatrix} 0 & 0 & 0 & 0 \\ 0 & 0 & 0 & 0 \\ 0 & 0 & 0 & 0 \\ 0 & 0 & 0 & a \end{bmatrix}.$$

The recovery of the solution for the independent set before the elimination of the single-point constraints is done by

$$u_n = \begin{bmatrix} u_f \\ Y_s \end{bmatrix}.$$

The u_n to u_g transformation via u_m was shown at the end of the previous chapter. The summary of the two reduction steps completed so far is

$$\begin{bmatrix} K_{gg} & & \\ & \begin{bmatrix} K_{nn} & \\ & [K_{ff}] \end{bmatrix} \end{bmatrix}.$$

4.6 Rigid body motion support

After executing above steps it is still possible that the K_{ff} stiffens matrix is singular and the structure remains a free body exhibiting a rigid body motion. Such a motion does not produce internal forces in the structure and should be restrained in order to solve the flexible body problem.

A common occurrence of this in static analysis is when the engineer forgets to constrain the structure to ground in some way. This results in six mechanisms, where the entire structure can move in a stress-free manner in the three translational and three rotational directions. A set of functions that describes the mechanism motions are called rigid body mode shapes.

Accordingly, the stiffness matrix is partitioned into restrained (r) and unrestrained (l) partitions and the static equilibrium is posed as

$$K_{ff} u_f = \begin{bmatrix} K_{rr} & K_{rl} \\ K_{lr} & K_{ll} \end{bmatrix} \begin{bmatrix} u_r \\ u_l \end{bmatrix} = \begin{bmatrix} F_r \\ F_l \end{bmatrix} = F_f.$$

The displacements of the supported partition by definition are zero, hence the second equation yields

$$K_{ll} u_l = F_l.$$

The stiffness matrix of this problem is now non-singular and may be solved for the internal deformations as

$$u_l = K_{ll}^{-1} F_l.$$

To verify that the support provided by the engineer is adequate, we turn to the first equation. Utilizing again the zero displacements the first equation yields

$$K_{rl} u_l = F_r.$$

Substituting the u_l displacements, we find the loads necessary to support the structure in terms of the active loads as

$$F_r = K_{rl} K_{ll}^{-1} F_l.$$

Introducing the rigid body transformation matrix of

$$G_{rl} = -K_{rl} K_{ll}^{-1}$$

and expanding it to the f partition size as

$$G_{fr} = \begin{bmatrix} I_{rr} \\ G_{rl}^T \end{bmatrix}$$

provides a way to measure the quality of the rigid body support. By transforming the stiffness matrix as

$$\overline{K}_{rr} = G_{fr}^T K_{ff} G_{fr}$$

we obtain the rigid body stiffness matrix that should be computationally zero. Measuring the magnitude of the Euclidean norm of the rigid body stiffness matrix and comparing to the norm of the supported partition provides a rigid body error ratio:

$$\epsilon_r = \frac{||\overline{K}_{rr}||}{||K_{rr}||}.$$

If this ratio is not small enough, the r partition is not adequately specified and the l set stiffness matrix is still singular. The summary of the three reduction steps is

$$\begin{bmatrix} K_{gg} & \\ & \begin{bmatrix} K_{nn} & \\ & \begin{bmatrix} K_{ff} & \\ & [K_{ll}] \end{bmatrix} \end{bmatrix} \end{bmatrix}.$$

4.7   Constraint augmentation approach

There is an alternative way of dealing with constraints in finite element computations and that is by augmenting them to the system of equations, instead of the elimination methods discussed in the prior sections.

As earlier, let us first consider the multi-point constraints. In Section 4.2 the partitioning of

$$K_{gg}u_g = \begin{bmatrix} \overline{K}_{nn} & K_{nm} \\ K_{mn} & K_{mm} \end{bmatrix} \begin{bmatrix} u_n \\ u_m \end{bmatrix} = \begin{bmatrix} \overline{F}_n \\ F_m \end{bmatrix} = F_g,$$

was the basis of the elimination and here it will serve as the basis of the augmentation. Recall the relationship between the dependent and independent degrees of freedom of the model:

$$u_m = G_{mn}u_n,$$

where G_{mn} is the constraint matrix. Let us augment the system of equations by the constraint matrix with the help of Lagrange multipliers as follows:

$$\begin{bmatrix} \overline{K}_{nn} & K_{nm} & G_{mn}^T \\ K_{mn} & K_{mm} & -I \\ G_{mn} & -I & 0 \end{bmatrix} \begin{bmatrix} u_n \\ u_m \\ \lambda_m \end{bmatrix} = \begin{bmatrix} \overline{F}_n \\ F_m \\ 0 \end{bmatrix},$$

or

$$\begin{bmatrix} K_{gg} & G_{mg}^T \\ G_{mg} & 0_{mm} \end{bmatrix} \begin{bmatrix} u_g \\ \lambda_m \end{bmatrix} = \begin{bmatrix} F_g \\ 0_m \end{bmatrix}.$$

The physical meaning of the Lagrange multipliers will be the reaction forces at the multi-point constraints, computed as

$$\lambda_m = K_{mn}u_n + K_{mm}u_m - F_m.$$

Similar approach to the single-point constraint partitioning form of

$$K_{nn}u_n = \begin{bmatrix} K_{ff} & K_{fs} \\ K_{sf} & K_{ss} \end{bmatrix} \begin{bmatrix} u_f \\ u_s \end{bmatrix} = \begin{bmatrix} F_f \\ F_s \end{bmatrix}$$

by augmenting with the single point constraints represented by the Y_s matrix results in

$$\begin{bmatrix} K_{ff} & K_{fs} & 0 \\ K_{sf} & K_{ss} & -I \\ 0 & I & 0 \end{bmatrix} \begin{bmatrix} u_f \\ u_s \\ \lambda_s \end{bmatrix} = \begin{bmatrix} F_f \\ F_s \\ Y_s \end{bmatrix},$$

or

$$\begin{bmatrix} K_{nn} & G_{sn}^T \\ G_{sn} & 0_{ss} \end{bmatrix} \begin{bmatrix} u_n \\ \lambda_s \end{bmatrix} = \begin{bmatrix} F_n \\ Y_s \end{bmatrix}.$$

The single point constraint forces are recovered as

$$\lambda_s = K_{sf} u_f + K_{ss} u_s - F_s.$$

The augmentation approach may be processed simultaneously by introducing two new partitions. First combine the two constraint sets into one as

$$2 = \begin{bmatrix} s \\ m \end{bmatrix}.$$

Then augment the g-partition with this partition to obtain the super-partition

$$1 = \begin{bmatrix} g \\ 2 \end{bmatrix}.$$

With these, the augmented form of the constrained linear static problem becomes

$$K_{11} u_1 = F_1,$$

where

$$F_1 = \begin{bmatrix} F_g \\ Y_2 \end{bmatrix}.$$

The enforced displacement term is a simple extension

$$Y_2 = \begin{bmatrix} Y_s \\ 0_m \end{bmatrix}.$$

The system matrix is

$$K_{11} = \begin{bmatrix} K_{gg} & G_{2g}^T \\ G_{2g} & 0 \end{bmatrix},$$

where the combined, single and multi-point constraint matrix is of form

$$G_{2g} = \begin{bmatrix} G_{sg} \\ G_{mg} \end{bmatrix}.$$

The G_{mg} sub-matrix was defined above as an appropriately scattered version of the G_{mn} sub-matrix, while the G_{sg} sub-matrix is based on the earlier G_{sn} matrix.

This augmented linear system may also be solved with specific pivoting techniques, such as described in [2]. These are needed due to the fact that the augmented stiffness matrix is indefinite on the account of the zero diagonal block introduced. Nevertheless, factorization techniques, such as discussed in Sections 7.2 and 7.3 solve these problems routinely.

The simultaneously augmented solution is partitioned as

$$u_1 = \begin{bmatrix} u_g \\ \lambda_2 \end{bmatrix},$$

where the first partition is the ultimate subject of the engineer's interest and the reaction forces at the constraints are trivial to compute.

The method of augmentation is not as wide-spread as the elimination, mainly because solver components were designed with positive-definite matrix focus in the past. The augmented approach is clearly more efficient for effects such as enforced motion, a subject of Section 15.5. It is also the basis of coupling finite element systems with multi-body analysis software, as shown in Section 11.6.

The next issue is the detection of singularities introduced by modeling techniques or errors in them. These may be of mechanisms, massless degrees of freedom, or even the dangerous combination of both, the massless mechanisms, subjects of the next chapter.

References

[1] Barlow, J.; Constraint relationships in linear and nonlinear finite element analyses, Int. Journal for Numerical Methods in Engineering, Vol. 2, Nos. 2/3, 149-156, 1982

[2] Bunch, J. R. and Parlett, B. N.; Direct methods for solving symmetric indefinite systems of linear equations, SIAM Journal of Numerical Analysis, Vol. 8, No. 2, 1971

[3] Komzsik, L.; The Lagrange multiplier approach for constraint processing in finite element applications, Proceedings of microCAD-SYSTEM '93, Vol. M, pp. 1-6, The University of Miskolc, Hungary, 1993

[4] Komzsik, L. and Chiang, K.-N.; The effect of a Lagrange multiplier approach on large scale parallel computations, Computing Systems in Engineering, Vol. 4, pp. 399-403, 1993

5

Singularity Detection of Finite Element Models

It is clear from the example of Chapter 4 that singularities may remain in the system after eliminating the specified constraints. The possible remaining singularities in the system may be due to modeling techniques. For example modeling planar behavior via membrane elements causes singularities due to the out of plane, drilling degrees of freedom being undefined. Similarly, in models with solid elements, the nodes have no rotational degrees of freedom. These are all called local singularities, the topic of Section 5.1.

There are also global singularities caused by rank deficiency of the stiffness matrix due to a part of the model having an unconstrained rigid body motion. These are called mechanisms and discussed in Section 5.2.

5.1 Local singularities

These singularities, once the proper degrees of freedom are identified, are removed by adding additional single-point constraints to the s partition and repeating the single-point elimination step.

In order to describe the detection process, let us consider the simple bar example again. The f partition version of the equilibrium equation of the model has only one nonzero term in a 4×4 matrix, so it is clearly singular.

The most practical technique to further identify singularities is by the systematic evaluation of node point singularity. This is executed by numerically examining the 3 by 3 sub-matrices corresponding to the translational and rotational degrees of freedom of every node. Let us denote the translational 3 by 3 sub-matrix of the i-th node by K_t^i and the rotational 3 by 3 sub-matrix by K_r^i. In case the K_{ff} matrix does not contain the full 3 by 3 sub-matrix of a certain node (due to constraint elimination), a smaller 2 by 2 or ultimately 1 by 1 sub-matrix is examined.

The process is based on the following singular value decompositions:

$$K_t^i = U_t \Sigma_t V_t^T,$$

and

$$K_r^i = U_r \Sigma_r V_r^T.$$

If any of the singular values obtained in Σ_r or Σ_t are less than a small computational threshold, then the corresponding direction defined by the appropriate column of V^T is considered to be singular. The singular direction may be resolved by an appropriate single-point constraint.

Naturally, more than one direction may be singular and ultimately all three may be singular. There may also be cases when a singular direction is not directly aligned with any of the coordinate axes. In this case, the coordinate direction closest to the singular vector (based on direction cosines) is constrained.

For example in the case of our example

$$K_r^1 = \begin{bmatrix} 0\ 0\ 0 \\ 0\ 0\ 0 \\ 0\ 0\ 0 \end{bmatrix}.$$

The singular value decomposition is trivial; all 3 directions are singular. The automatically found singularities are:

$$u_s^a = \begin{bmatrix} \theta_x(1) \\ \theta_y(1) \\ \theta_z(1) \end{bmatrix}.$$

The superscript notes that this set is the automatically detected singular set. The f-size equilibrium equation may now be reduced by applying the automatic single point constraints. The automatic single-point constraints are described by

$$u_s^a = Y_s^a,$$

where the u_s^a is a partition of u_f:

$$u_f = \begin{bmatrix} u_a \\ u_s^a \end{bmatrix}.$$

It follows, that for our example:

$$Y_s^a = \begin{bmatrix} 0.0 \\ 0.0 \\ 0.0 \end{bmatrix}.$$

Applying these to the f-size equilibrium equation results in

$$K_{aa}u_a = F_a - K_{as^a}Y_s^a.$$

This is commonly called the analysis set formulation as after this step, the numerical model is ready for analysis. The following chart summarizes all the numerical reduction steps leading to the analysis formulation

$$\begin{bmatrix} K_{gg} \\ \quad \begin{bmatrix} K_{nn} \\ \quad \begin{bmatrix} K_{ff} \\ \quad \begin{bmatrix} K_{ll} \\ \quad \begin{bmatrix} K_{aa} \end{bmatrix} \end{bmatrix} \end{bmatrix} \end{bmatrix} \end{bmatrix}.$$

The final stiffness matrix ready for analysis is

$$K_{aa} = \begin{bmatrix} a \end{bmatrix},$$

and the final, clearly non-singular, equilibrium equation is

$$K_{aa}u_a = F_a,$$

or

$$aq_x(2) = F.$$

Assuming numerical values of $E = 10^7, A = 0.1, F = 1.0$ in the appropriate matching units, the numerical solution of our example problem is a displacement of 10^{-6} units. Figure 5.1 shows the (exaggerated) deformed shape of the example model.

  The recovery of the solution prior to the single-point constraint elimination is

$$u_n = \begin{bmatrix} 0.0 \\ 0.0 \\ 0.0 \\ 0.0 \\ 0.0 \\ 0.0 \\ 10^{-6} \end{bmatrix} = \begin{bmatrix} q_x(1) \\ q_y(1) \\ q_z(1) \\ \theta_x(1) \\ \theta_y(1) \\ \theta_z(1) \\ q_x(2) \end{bmatrix}.$$

Here the first 3 rows containing zeroes are the boundary condition Y_s constraints. The second 3 represent the automatically found Y_s^a constraints. This vector is then further reprocessed to reflect the multi-point constraint elimination.

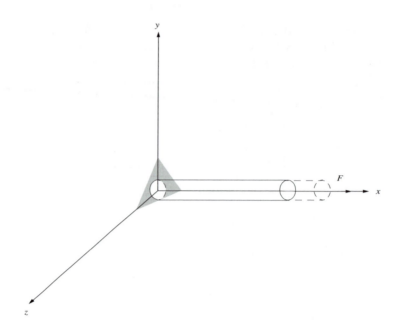

FIGURE 5.1 Deformed shape of bar

$$u_m = G_{mn} u_n = \begin{bmatrix} 0.0 \\ 0.0 \\ 0.0 \\ 0.0 \\ 0.0 \end{bmatrix} = \begin{bmatrix} q_y(2) \\ q_z(2) \\ \theta_x(2) \\ \theta_y(2) \\ \theta_z(2) \end{bmatrix}.$$

The zero result is explained by the fact that the last column of the G_{mn} matrix derived in Section 4.2 is zero and only the last term of the u_n vector is nonzero. The g partition result vector is obtained by merging the m and n partitions:

$$
u_g = \begin{bmatrix} 0.0 \\ 0.0 \\ 0.0 \\ 0.0 \\ 0.0 \\ 0.0 \\ 10^{-6} \\ 0.0 \\ 0.0 \\ 0.0 \\ 0.0 \\ 0.0 \end{bmatrix} = \begin{bmatrix} q_x(1) \\ q_y(1) \\ q_z(1) \\ \theta_x(1) \\ \theta_y(1) \\ \theta_z(1) \\ q_x(2) \\ q_y(2) \\ q_z(2) \\ \theta_x(2) \\ \theta_y(2) \\ \theta_z(2) \end{bmatrix}.
$$

This is the final result of our computational example; the displacement at every node point and each degree of freedom is known.

5.2 Global singularities

Global singularities occur when the stiffness matrix is rank deficient due to linear dependency among the columns. They manifest an engineering phenomenon called a mechanism. After all flexible elements have been accounted for, the constrained degrees of freedom have been eliminated and rigid body motions are supported, it is still possible to have groups of degrees of freedom that are free to move without incurring loads in other degrees of freedom.

For example, an open automobile door is free to rotate without causing loads in the hinges and other structures to which it attaches. Each group of degrees of freedom which can move in a stress-free manner are attached to what is called a mechanism. The presence of mechanisms causes difficulties in the solution process unless they are constrained by some other device. Degrees of freedom on a mechanism are likely to move through a large motion for moderate loads, and a small change in loads can lead to a large change in response.

The above car door example is a local mechanism. If it is a part of the model where constraints to ground were neglected the model has seven mechanisms. This condition is sometimes described as seven unconstrained rigid body modes being present in the model. Aerospace industry models may have even larger numbers of rigid body modes due to a larger number of door-like components. The flaps and other stability control components all could represent mechanisms.

Mechanisms may or may not have been intentional by the engineer. In any case these type of singularities also need to be removed before any numerical solution is attempted. After eliminating the constraints and the local singularities, this situation could still happen if there is a linear dependency among the columns of K_{aa}. To recognize this case, we rely on some diagnostics provided during the

$$K_{aa} = L_{aa} D_{aa} L_{aa}^T$$

factorization (more details on the process in the next chapter) of the stiffness matrix as follows.

Due to round-off, mechanisms usually manifest themselves as very small values of $D(i,i)$. That of course could result in an unstable factorization and unreliable solutions later. To prevent this, the ratio of

$$r(i) = K_{aa}(i,i)/D_{aa}(i,i)$$

is monitored throughout the factorization. This strategy is based on the recommendation of [2] and used widely in the industry where it is called the matrix/factor diagonal ratio. Wherever this ratio exceeds a certain threshold the related degree of freedom is considered to be part of a possible mechanism. The threshold used in practice is usually close to $\sqrt{\epsilon_{machine}}$, where $\epsilon_{machine}$ is representative of the floating point arithmetic accuracy of the computer used.

In large models the straightforward listing of these degrees of freedom may not be adequate to aid the engineer in resolving the problem. To visualize the "shape" of the mechanism, let us partition the degrees of freedom into a partition (x) exceeding and the complementary set $(\bar{x})$ below the threshold.

$$u_a = \begin{bmatrix} u_{\bar{x}} \\ u_x \end{bmatrix}.$$

A similar partitioning of the stiffness matrix is

$$K_{aa} = \begin{bmatrix} K_{\bar{x}\bar{x}} & K_{\bar{x}x} \\ K_{x\bar{x}} & K_{xx} \end{bmatrix}.$$

Let us virtually apply a force (P_x) to the degrees of freedom with exceedingly high ratios. Let us assume that this virtual force moves the degrees of freedom in the x partition by one unit.

$$\begin{bmatrix} K_{\bar{x}\bar{x}} & K_{\bar{x}x} \\ K_{x\bar{x}} & K_{xx} \end{bmatrix} \begin{bmatrix} u_{\bar{x}} \\ I_x \end{bmatrix} = \begin{bmatrix} 0 \\ P_x \end{bmatrix}.$$

The solution shape of the stable, below the threshold partition is

$$u_{\bar{x}} = -K_{\bar{x}\bar{x}}^{-1} K_{\bar{x}x}.$$

This shape helps the analyst to recognize the components of the mechanism. The drawback of this method of identifying mechanisms is its non-trivial cost, therefore, it is executed only at the engineer's specific request in commercial software.

The solution of the mechanism problem is to tie the independent component to the rest of the structure. This is done by applying more flexible elements and/or appropriate multi-point constraints and repeating the multi-point elimination process. Writing these constraint equations is not trivial and needs engineering knowledge in most cases.

5.3   Massless degrees of freedom

So far our focus has been restricted to the stiffness matrix. The mass matrix whose generation was shown in Chapter 3 usually has a large number of zero rows and columns. In industrial finite element analysis these are called massless degrees of freedom. Some of the eigenvalue analysis methods on the other hand require a nonsingular mass matrix. Such method is the dense matrix reduction type method of Householder, to be further discussed in Chapter 9.

In order to allow the use of such methods, another reduction step is sometimes executed in the industry, to eliminate the massless degrees of freedom from both the global mass and the global stiffness matrix.

Let us consider the a partition normal modes analysis problem of

$$[K_{aa} - \lambda M_{aa}]u_a = 0.$$

Remember, this problem now has the global singularities also removed. Let these matrices be partitioned into the o partition of massless degrees of freedom and the $\bar{a}$ partition of degrees of freedom with masses.

$$\begin{bmatrix} K_{oo} & K_{o\bar{a}} \\ K_{o\bar{a}}^T & K_{\bar{a}\bar{a}} - \lambda M_{\bar{a}\bar{a}} \end{bmatrix} \begin{bmatrix} u_o \\ u_{\bar{a}} \end{bmatrix} = 0.$$

The fact that the mass matrix contains zero rows and columns in the o partitions is demonstrated by the lack of M terms in the appropriate partitions. Introducing a transformation matrix

$$G_{a\bar{a}} = \begin{bmatrix} -K_{oo}^{-1}K_{o\bar{a}} \\ I_{\bar{a}\bar{a}} \end{bmatrix},$$

the displacement vector is reduced as

$$u_a = G_{a\bar{a}} u_{\bar{a}}.$$

Left multiplication by the transformation matrix yields

$$G_{a\bar{a}}^T (K_{aa} - \lambda M_a a) G_{a\bar{a}} u_{\bar{a}} = 0.$$

With

$$K_{\bar{a}\bar{a}} = G_{a\bar{a}}^T K_{aa} G_{a\bar{a}},$$

and

$$M_{\bar{a}\bar{a}} = G_{a\bar{a}}^T M_{aa} G_{a\bar{a}},$$

one obtains a reduced set of equations devoid of massless degrees of freedom

$$[K_{\bar{a}\bar{a}} - \lambda M_{\bar{a}\bar{a}}] u_{\bar{a}} = 0.$$

5.4   Massless mechanisms

Massless mechanisms are the most dangerous and annoying singularities in finite element computations. They are a combination of the mechanisms and massless degrees of freedom, occurring when the stiffness and mass matrices have a coinciding zero subspace.

Recall that the rigid body eigenvalues were computational zeroes. The massless degrees of freedom, if left in the system, may result in infinite eigenvalues. Neither of them is troublesome on its own right, but their combination is especially troublesome because the eigenvalues corresponding to that case are indefinite. This may carry the grave consequence of finding spurious modes.

The practical scenario, under which this may happen, is when either some mass or stiffness components are left out of the model due to engineering error. It is very important to detect such cases and this may be done as follows. Let us compute a linear combination of the stiffness and mass matrix as

$$A = K_{\bar{a}\bar{a}} + \lambda_s M_{\bar{a}\bar{a}}.$$

Note that the combination coefficient is positive, resulting in a negative shift in an eigenvalue analysis sense, avoiding the influence of the eigenvalues to the left of the shift.

The detection is again based on the factorization

$$A = L_{AA}D_{AA}L_{AA}^T,$$

and the ratio of matrix diagonals and factor diagonals is monitored:

$$R(i) = A(i,i)/D_{AA}(i,i).$$

The following strategy is based on the industrially acknowledged solution in [3] and may be used to correct massless mechanism scenarios. Based on the R vector a P matrix originally initialized to zero is populated as follows. For $i = 1, 2, ...n$, if

$$R(i) > threshold,$$

then the ith row of the next available column $j, j = 1, 2, ..m$ of the P matrix will be set to unity as

$$P(i,j) = 1.$$

The threshold is the maximum ratio, for example 10^6, tolerated by the engineer. The process results in having only one nonzero term in each column of the P matrix. If there is no entry in R that violates the maximum threshold, then P is empty and there are no massless mechanisms detected in the system.

The solution of the system

$$AU = P$$

by exploiting the factorization executed for the detection

$$U = (L_{AA}D_{AA}L_{AA}^T)^{-1}P$$

provides the shapes depicting the massless mechanism. Each column of the U matrix represents a potential massless mechanism mode, assuming that the rigid body modes were already removed.

The locations at the end of each of the remaining mode shapes are gathered into a single partitioning vector Q. The nonzero terms in the vector constitute the $\overline{m}$ partition of massless mechanism degrees of freedom. The vector is used to partition both the stiffness and mass matrices simultaneously.

$$K_{\overline{aa}} = \begin{bmatrix} K_{\overline{aa}} & K_{\overline{am}} \\ K_{\overline{ma}} & K_{\overline{mm}} \end{bmatrix}$$

and

$$M_{\overline{aa}} = \begin{bmatrix} M_{\overline{aa}} & M_{\overline{am}} \\ M_{\overline{ma}} & M_{\overline{mm}} \end{bmatrix}.$$

The $\overline{m}$ partition represents the subspaces corresponding to the massless mechanisms and as such, discarded. The pencil of

$$(K_{\overline{\overline{aa}}}, M_{\overline{\overline{aa}}})$$

may be now safely subjected to eigenvalue analysis. The final reduction summary chart after the elimination of massless degrees of freedom and mechanisms is

$$\begin{bmatrix} K_{gg} & \\ & \begin{bmatrix} K_{nn} & \\ & \begin{bmatrix} K_{ff} & \\ & \begin{bmatrix} K_{ll} & \\ & \begin{bmatrix} K_{aa} & \\ & \begin{bmatrix} K_{\overline{aa}} & \\ & \begin{bmatrix} K_{\overline{\overline{aa}}} \end{bmatrix} \end{bmatrix} \end{bmatrix} \end{bmatrix} \end{bmatrix} \end{bmatrix} \end{bmatrix}.$$

There are further singularity phenomena, such as for example singularities at element corner nodes [1]. These are not addressed via generic reduction approaches, but by specific numerical adjustments, and as such, not discussed here further. It is also possible that multiple physical phenomena are modeled simultaneously, for example elasticity with fluid dynamics, and this topic will be discussed in the next, final chapter of Part I, after reviewing some industrial case studies.

5.5   Industrial case studies

To quantitatively demonstrate the effect of the numerical finite element model generation techniques discussed so far, several automobile industry examples were collected. Table 5.1 shows the number of nodes and various elements of three finite element models. They were automobile body examples such as shown in Figure 5.2.

Such structural models are called "body-in-white" models in the automobile industry to distinguish from the fully equipped "trimmed" models shown in Chapter 9. They all had various shell and solid elements. The shell elements include 4 or 8 noded quadrilateral elements and 3 or 6 noded triangular elements (first and second order elements, respectively). The solid elements were hexahedral, pentahedral (also called wedge) and 4 or 10 noded (first or second order) tetrahedral elements.

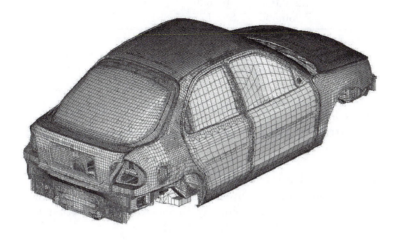

FIGURE 5.2 Typical automobile body-in-white model

TABLE 5.1
Element statistics of automobile model examples

Model	Nodes	Shells	Solids	Rigids
A	529,116	428,154	38,385	59,071
B	712,680	563,563	123,602	50,868
C	1,322,766	1,195,701	89,980	4,766

The matrix partition sizes corresponding to the various steps of numerical model generation are collated in Table 5.2. The g partition is the global, assembled partition, the n partition is what remains after eliminating the multi-point constraints. The f partition has the single-point constraints also removed and finally the a partition, ready for analysis, has no singularities.

It is important to notice that the difference between the n and the f partitions is rather small, indicating the practical industrial tendency of engineers giving only a few single-point constraints (as boundary conditions) explicitly.

TABLE 5.2
Reduction sizes of automobile model examples

Model	g	n	f	a
A	3,135,144	2,857,100	2,856,080	2,643,561
B	4,276,080	4,101,639	4,101,557	3,577,998
C	7,936,560	7,882,177	7,881,784	6,936,560

There is much reliance on automated singularity processing, demonstrated by the difference between the f and a partitions.

On the other hand, the large number of rigid elements gives rise to many multi-point constraints. This produces the noticeably large difference between the g and the n partition sizes.

We have now arrived at the numerical model containing matrices ready for various analyses. Due to the often enormous size of this model, a variety of reduction methods is used in the industry to achieve a computational model. These are the subject of the chapters of Part II.

References

[1] Benzley, S. E,; Representation of singularities with iso-parametric finite elements, Numerical Methods in Engineering, Vol. 8, No. 3, pp. 537-545, 2005

[2] Gockel, M. A.; An index for the quality of linear equation solutions, Lockheed Corp. Technical Report, LR 25507, 1973

[3] Gockel, M. A.; Massless mechanism detection for real modes, MSC User report MMA.v707, 1999

6

Coupling Physical Phenomena

The life cycle of products contains scenarios when the structure is interacting with another physical entity. Important practical problems are, for example, when the structure surrounds or immersed into a volume of fluid, both called fluid-structure interaction scenarios.

6.1 Fluid-structure interaction

We focus on analysis scenarios when the behavior is dominated by the structure. The coupling between the fluid and the structure introduces an unsymmetric matrix component. The coupled equilibrium of an undamped structure is

$$M_c \ddot{u}_c + K_c u_c = F_c,$$

where the subscript c refers to the coupling. The coupled matrices are

$$M_c = \begin{bmatrix} M_s & 0 \\ A & M_f \end{bmatrix}$$

and

$$K_c = \begin{bmatrix} K_s & -A^T \\ 0 & K_f \end{bmatrix}.$$

The subscripts s and f refer to the structure and fluid, respectively, not to the s of f partitions presented in prior sections. The A matrix represents the fluid-structure coupling and will be discussed shortly. The components of the coupled solution vector are

$$u_c = \begin{bmatrix} u_s \\ p_f \end{bmatrix},$$

where the p_f is the pressure in the fluid and u_s is the displacement in the structural part. The external load may also be given as structural forces and pressure values as

$$F_c = \begin{bmatrix} F_s \\ P_f \end{bmatrix}.$$

6.2   A hexahedral finite element

The fluid volume may be best represented by hexahedral finite elements. This enables the interior to be modeled by element shapes close to the Cartesian discretization of the volume. The element shapes are gradually deformed to adhere to the shape of the boundary of the fluid volume.

Following the principles laid down regarding the three dimensional continuum modeling by tetrahedral elements in Section 3.4, every node point of the linear (so-called eight-noded) hexahedron has three degrees of freedom and there are twenty four nodal displacements of the element.

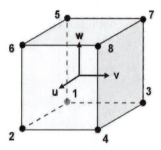

FIGURE 6.1 Hexahedral finite element

There are higher order hexahedral elements, for example having nodes in the middle of every edge will result in the so-called twenty-noded element. Finally, having nodes on the middle of faces as well as one in the middle of the volume produces the 27-noded element. We will review the 8-noded element such as shown in Figure 6.1 in the following.

The cube has sides of two units and the local coordinate system is originated in the center of the volume of the element, resulting in the local coordinate values of the node points described in Table 6.1.

TABLE 6.1

Local coordinates of hexahedral element

Node	u	v	w
1	-1	-1	-1
2	1	-1	-1
3	-1	1	-1
4	1	1	-1
5	-1	-1	1
6	1	-1	1
7	-1	1	1
8	1	1	1

This arrangement of the nodes corresponds to the quadrilateral element introduced in section 1.7 and consequently the shape functions of the hexahedral element may also be generalized for $i = 1, 2, \ldots, 8$ as

$$N_i = \frac{1}{8}(1 + u_i u)(1 + v_i v)(1 + w_i w).$$

Developing the formula for $i = 1$ as an example yields

$$N_1 = \frac{1}{8}(1 - u)(1 - v)(1 - w),$$

which is an obvious generalization of the quadrilateral element's shape function. The vector of nodal displacements for the element becomes

$$q_e = \begin{bmatrix} q_{1x} \\ q_{1y} \\ q_{1z} \\ \dots \\ \dots \\ \dots \\ q_{8x} \\ q_{8y} \\ q_{8z} \end{bmatrix}.$$

Recall that q_{ix} refers to the x translation of the ith local node of the element. The displacement at any location inside this element is approximated as

$$q(x, y, z) = N q_e.$$

The organization of the N matrix of the eight shape functions is

$$N = \begin{bmatrix} N_1 & 0 & 0 & N_2 & 0 & 0 & \dots & N_8 & 0 & 0 \\ 0 & N_1 & 0 & 0 & N_2 & 0 & \dots & 0 & N_8 & 0 \\ 0 & 0 & N_1 & 0 & 0 & N_2 & \dots & 0 & 0 & N_8 \end{bmatrix}.$$

The location of a point inside the element is approximated again with the same four shape functions as the displacement field

$$x = \sum_{i=1}^{8} N_i x_i,$$

$$y = \sum_{i=1}^{8} N_i y_i,$$

and

$$z = \sum_{i=1}^{8} N_i z_i.$$

Here x_i, y_i, z_i are the eight x, y, z coordinates of the nodes of the hexahedron, hence the element is again an iso-parametric element.

Further mimicking of the procedures developed for the quadrilateral element (Section 1.7) and the tetrahedral element (Section 3.4) results in the element stiffness matrix of

$$k_e = \int_{-1}^{+1} \int_{-1}^{+1} \int_{-1}^{+1} B^T D B |J| \, du \, dv \, dw,$$

where B and J are computed before in the sections mentioned above.

6.3   Fluid finite elements

The physical behavior of the fluid is governed by the wave equation of form:

$$\frac{1}{b}\ddot{p} = \nabla(\frac{1}{\rho}\nabla p).$$

Here

$$b = c^2\rho_0$$

is the bulk modulus, c is the speed of sound in the fluid and ρ is its density [4]. The connection with the surrounding structure is based on relating the structural displacements to the pressure in the fluid.

The boundary condition at a structure-fluid interface is defined as

$$\frac{\partial p}{\partial \underline{n}} = -\rho\ddot{u}_{\underline{n}},$$

where $\underline{n}$ is the direction of the outward normal. At free surfaces of the fluid we assume

$$u = p = 0.$$

We address the problem in its variational formulation [3]

$$\int\int\int_V [\frac{1}{b}\ddot{p} - \frac{1}{\rho}\nabla\cdot\nabla p]p\,dV = 0.$$

The finite element discretization of this can be carried out again based on Galerkin's principle assuming

$$p(x,y,z) = \sum_{i=1}^{n} N_i p_i = N\underline{p},$$

where p_i and N_i are the discrete pressure value and shape functions associated with the i-th node of the fluid finite element mesh.

The same holds for the derivatives:

$$\ddot{p}(x,y,z) = N\underline{\ddot{p}}.$$

Separating the two parts of the variational equation, the first yields

$$\int_V \frac{1}{b}\ddot{p}p\,dV = \int_V \frac{1}{b}p\ddot{p}\,dV = \underline{p}^T \int_V \frac{1}{b}N^T N\,dV\underline{\ddot{p}}.$$

Introducing the fluid mass matrix

$$M_f = \int_V \frac{1}{b} N^T N dV,$$

this term simplifies to

$$\int_V \frac{1}{b} \ddot{p} p dV = \underline{p}^T M_f \underline{\ddot{p}}.$$

The components of the fluid mass matrix are computed as

$$M_f(i,j) = \frac{1}{b} \int_V N_i N_j dV.$$

The second part of the variational equation integrated by parts yields

$$-\int_V (\frac{1}{\rho} \nabla \cdot \nabla p p) dV = \int_V \frac{1}{\rho} \nabla p \cdot \nabla p dV - \int_S \frac{1}{\rho} \nabla p p dS.$$

From above assumptions it follows that

$$\nabla p = \nabla N \underline{p},$$

and using the boundary condition stated above, we obtain

$$\underline{p}^T \int_V (\frac{1}{\rho} \nabla N^T) \nabla N dV \, \underline{p} + \underline{p}^T \int_S N^T \ddot{u}_n dS.$$

Introducing the fluid stiffness matrix of form

$$K_f = \int_V \frac{1}{\rho} \nabla N^T \nabla N dV,$$

the first part of the variational equation simplifies to

$$\underline{p}^T K_f \underline{p}.$$

The components of the fluid stiffness matrix are computed as

$$K_f(i,j) = \frac{1}{\rho} \int_V \nabla N_i \nabla N_j dV.$$

The coupling force exerted on the boundary by the surrounding structure is

$$A = \int_S N^T \ddot{u}_n dS.$$

Utilizing the boundary condition, the terms of the coupling matrix are computed as

$$A(i,j) = \int N_i N_j dS.$$

6.4   Coupling structure with compressible fluid

A symmetric coupled formulation is also possible using the same constituent matrices but with a different order of operations. It produces the same solution but with symmetric matrices, which can be solved more economically and reliably.

This method is applicable when the fluid component is highly compressible, as it requires the inverse of the fluid matrices. Since these matrices are usually significantly smaller than the structural matrices, the cost of obtaining these inverses is not prohibitive.

The derivation is as follows. Let us write out the first row of the coupled matrix equation

$$M_{ss}\ddot{u} + K_{ss}u - A^T p = F_s.$$

Add and subtract a yet rather cryptic term as

$$M_{ss}\ddot{u} + K_{ss}u - A^T p + A^T M_{ff}^{-1} Au - A^T M_{ff}^{-1} Au = F_s.$$

After grouping and reorganizing

$$M_{ss}\ddot{u} + (K_{ss} + A^T M_{ff}^{-1} A)u - A^T(p + M_{ff}^{-1} Au) = F_s.$$

This will become the first equation of the new coupled form. The second coupled equation is written as

$$A\ddot{u} + M_{ff}\ddot{p} + K_{ff}p = F_f.$$

Pre-multiplying by $M_{ff}K_{ff}^{-1}$ and again adding and subtracting a specifically chosen term results in

$$M_{ff}K_{ff}^{-1} A\ddot{u} + M_{ff}K_{ff}^{-1} M_{ff}\ddot{p} + M_{ff}K_{ff}^{-1} K_{ff}p + Au - Au = M_{ff}K_{ff}^{-1} F_f.$$

Grouping and reordering again produces the second equation of the new coupled form

$$M_{ff}K_{ff}^{-1} M_{ff}(\ddot{p} + M_{ff}^{-1} A\ddot{u}) - Au + M_{ff}(p + M_{ff}^{-1} Au) = M_{ff}K_{ff}^{-1} F_f.$$

Introducing the new variable

$$q = p + M_{ff}^{-1} Au$$

and its 2nd derivative

$$\ddot{q} = \ddot{p} + M_{ff}^{-1} A\ddot{u}$$

results in the symmetric coupled formulation of

$$\begin{bmatrix} M_{ss} & 0 \\ 0 & M_{ff}K_{ff}^{-1}M_{ff} \end{bmatrix}\begin{bmatrix} \ddot{u} \\ \ddot{q} \end{bmatrix} + \begin{bmatrix} K_{ss} + A^T M_{ff}^{-1}A & -A^T \\ -A & M_{ff} \end{bmatrix}\begin{bmatrix} u \\ q \end{bmatrix} = \begin{bmatrix} F_s \\ M_{ff}K_{ff}^{-1}F_f \end{bmatrix}.$$

More details on the physics and another alternative formulation may be seen in [1].

6.5   Coupling structure with incompressible fluid

Coupling structures with incompressible fluid results in a simpler computational form that is still based on the coupled equilibrium as

$$M_c\ddot{u}_c + K_c u_c = F_c,$$

however, due to the incompressibility of the fluid there is no acceleration dependent fluid term in the coupled mass matrix:

$$M_c = \begin{bmatrix} M_s & 0 \\ A & 0 \end{bmatrix}.$$

The stiffness matrix remains

$$K_c = \begin{bmatrix} K_s & -A^T \\ 0 & K_f \end{bmatrix}.$$

This enables the elimination of the pressure from the second equation,

$$A\ddot{u} + K_f p = 0,$$

resulting in

$$p = -K_f^{-1}A\ddot{u}.$$

Substituting into the first equation produces

$$M_s\ddot{u} + K_s u - A^T p = M_s\ddot{u} + K_s u + A^T K_f^{-1}A\ddot{u}.$$

By introducing the so-called virtual mass

$$M_v = A^T K_f^{-1}A,$$

the coupled equations of motion simplify to

$$(M_s + M_v)\ddot{u} + K_s u = 0.$$

FIGURE 6.2 Fuel tank model

This computation is very practical for fluid containers, such as the fuel tank of automobiles shown in Figure 6.2.

Other applications include structures surrounded by water, such as ships and off-shore drilling platforms.

6.6 Structural acoustic case study

In the interior acoustics application of car bodies [2] the interior fluid is the air, and as such it is compressible. The aim of acoustic response analysis is to compute and ultimately reduce the noise at a certain location inside of an automobile or a truck cabin shown in Figure 6.3.

An example from a leading auto-maker had 5.6 million total coupled degrees of freedom. The structural excitation originated from the front tires and

FIGURE 6.3 Truck cabin model

the goal was to compute the pressure (the noise) at the hypothetical driver's ear.

Table 6.2 contains statistics of the matrices in this analysis. Note the very large number of the zero columns in the mass matrix resulting from the fact that the external body model is mostly from shell elements. The rather higher density of the stiffness matrix, manifested by the more than ten-thousand nonzero terms in some columns, is also noteworthy. This is a result of the presence of the interior fluid model build from solid elements with higher connectivity.

TABLE 6.2
Acoustic response analysis matrix statistics

K	number of rows	nonzero terms	max terms
matrix	4.25 million	156 million	10,322

M	number of rows	zero columns	max terms
matrix	4.25 million	665,486	12

The analysis was executed on a workstation with 4 (1.7 GHz) CPUs. The overall analysis required 1,995 elapsed minutes, of which about 840 minutes was the eigenvalue solution. The amount of I/O executed was 4.3 Terabytes and the disk footprint was 183 Gigabytes.

Note that a series of such forced vibration problems are solved in practice, between which the structural model is changed to modify the acoustic response (lower the noise). These computations are very time consuming, therefore usage of advanced computational methods in the solution components, subject of Part II are of paramount importance.

Furthermore, this physical problem may also contain damping and could result in a generalized quadratic eigenvalue problem (zero excitation) or forced, damped vibration (nonzero excitation), topics also discussed at length in Part II.

References

[1] Everstine, G. C.; Finite element formulation of structural acoustics problems, Computers and Structures, Vol. 65, No. 3, pp. 307-321, 1997

[2] Komzsik, L.; Computational acoustic analysis in the automobile industry, Proc. Supercomputer Applications in the Automobile Industry, Florence, Italy, 1996

[3] Komzsik, L.; Applied variational analysis for engineers, Taylor and Francis, Boca Raton, 2009

[4] Warsi, Z. U. A.; Fluid dynamics, Taylor and Francis, Boca Raton, 2006

Part II

Computational Reduction Techniques

7

Matrix Factorization and Linear Systems

Factorization of the finite element matrices and the solution of linear systems plays a significant role in the following chapters. They are also the foundation of one of the practical problems mentioned earlier, the linear static analysis. Therefore, we focus this chapter on these topics.

7.1 Finite element matrix reordering

The main reasons for reordering the finite element matrices are to minimize:

1. the storage requirements of the factor matrix,
2. the computing time of the factorization, and
3. the round-off errors during the computation.

Reordering means replacing various rows and the corresponding columns of the matrix while maintaining symmetry. To demonstrate this, we look at the following two matrices:

$$
A = \begin{bmatrix}
x & x & x & x & & & & \\
x & x & & & & & & \\
x & & x & & x & & & \\
x & & & x & & x & & \\
& & & & x & & & \\
& & & & & x & & \\
& & & & & & x & \\
& & & & & & & x
\end{bmatrix}
$$

and

$$B = \begin{bmatrix} x & & & & & & & & \\ & x & & & & & & & \\ & & x & & & & & & \\ & & & x & & & & & \\ & & & & x & & x & & \\ & & & & & x & x & & \\ & & & & & & x & x & \\ & & & & & x & x & x & x \end{bmatrix}.$$

The x locations are the nonzero values of the matrices and the other locations hold zeroes. As noted in earlier chapters, the assembled finite element matrices are very sparse. It is easy to see that the two matrices are reordered versions of each other.

Specifically, the last four rows and columns of the A matrix were "reordered" to be the the first four rows and columns of the B matrix. Furthermore the first row and column of A became the last row and column of B. This choice, and the "betterness" of the B matrix becomes obvious when viewing the factorization of

$$A = LDL^T.$$

Here the L matrix is a unit triangular matrix and D is diagonal. An algorithm for such a factorization of an order n symmetric matrix may be written as:

For $i = 1, n$

$$D(i, i) = A(i, i) - \Sigma_{k=1}^{i-1} L(i, k)^2 D(k, k)$$

For $j = i + 1, n$

$$L(j, i) = (A(j, i) - \Sigma_{k=1}^{i-1} L(i, k)D(k, k)L(j, k))/D(i, i)$$

End loop j
End loop i

It is important to notice that the inner loop involves a division by $D(i, i)$. If this term is zero, the matrix is singular and cannot be factored. Alternatively, if this term is very small, the round-off error of the division is very large. This is the third reason for reordering: moving small terms off the diagonal improves accuracy.

Executing the above factorization for both A and B matrices, the triangular factors have the following sparsity patterns.

$$L_A^T = \begin{bmatrix} x & x & x & x & & & & \\ & x & y & y & & & & \\ & & x & y & & & & \\ & & & x & & & & \\ & & & & x & & & \\ & & & & & x & & \\ & & & & & & x & \\ & & & & & & & x \end{bmatrix}$$

and

$$L_B^T = \begin{bmatrix} x & & & & & & & \\ & x & & & & & & \\ & & x & & & & & \\ & & & x & & & & \\ & & & & x & & x & \\ & & & & x & x & & \\ & & & & & x & x & \\ & & & & & & & x \end{bmatrix}.$$

The observation is that the L_A^T factor has new (fill-in) terms noted by y. These terms will increase the storage requirement of the factor (reason 1) and of course the computational time (reason 2).

Mathematically the reordering is presented by a permutation matrix, containing only zeroes and ones in specific locations. For our specific example

$$B = PAP^T$$

where

$$P = \begin{bmatrix} & & & & & & & 1 \\ & & & & & & 1 & \\ & & & & & 1 & & \\ & & & & 1 & & & \\ & & & 1 & & & & \\ & & 1 & & & & & \\ & 1 & & & & & & \\ 1 & & & & & & & \end{bmatrix}.$$

To calculate the P permutation matrix, it is best to consider a graph representation of the matrix A. We introduce the following equivalency criteria:

1. Diagonal terms of the matrix correspond to a vertex of the graph,
2. Off-diagonal terms of a matrix correspond to edges in the graph.

With these, every symmetric matrix has an equivalent undirected graph. Traversing the graph in a suitable manner yields a proper permutation matrix.

One practical traversal method is called the minimum degree algorithm [6]. The algorithm systematically removes nodes from the graph until the graph is eliminated. The selection of the next node is based on the node's connectivity (the number of other nodes that are connected). The node with the minimum connectivity will be removed, hence the name. Based on the the matrix-graph analogy, this means factoring the row of the matrix first with the least number of off-diagonal terms.

The process maintains adjacency of nodes by introducing temporary edges. The temporary edges correspond to the fill-in terms of the factorization process. The minimum degree elimination of the graph is the symbolic factorization of the corresponding matrix. Capturing the order of the elimination produces the P permutation matrix.

There are many advanced variations of the minimum degree method. They are reviewed in [3]. These methods are the predecessors of the domain decomposition methods mentioned in later chapters.

Note again, that the above deals with symmetric matrices. For unsymmetric matrices a factorization of

$$A = LU$$

is applied. The reordering is in this case represented by two distinct (row and column) permutation matrices as

$$PAQ = B.$$

Otherwise most of the discussion of symmetric matrices applies to unsymmetric matrices as well.

7.2 Sparse matrix factorization

The finite element method gives rise to very sparse matrices. This was especially obvious in our simple example in Chapter 3. Depending on the model and the discretization technique, matrices with less than a tenth of a percent density are commonplace. It is a natural desire to exploit this level of sparsity in the matrices.

First and foremost, if the matrices were stored as two-dimensional arrays, for today's multi-million degree of freedom problems the memory requirement would be in the terabytes. Secondly, the amount of numerical work executed on zero operations would be the dominant part of the factorization. Most of the computational techniques described in the remainder of the book could be implemented by taking the sparsity into consideration. We review the technique here briefly.

The factorization of finite element matrices is usually executed in two phases. The first phase is the symbolic phase, called such because the factorization process is executed by considering the location of the terms only. The second, numeric phase, executes the actual factorization on the terms.

The symbolic phase separates the topology structure of the matrix from its numerical content. In this form the matrix is stored in three linear arrays. One contains the row and another the column indices of every term, while a third one contains the numeric values. The matrix is then described as

$$A = [IA(k), JA(k), NA(k) \; ; \; k = 1, ...NZ],$$

where NZ is the number of nonzero terms in the matrix. Here IA and JA are integer arrays and NA is real. Specifically, a term $A(i,j)$ of the matrix is nonzero if there is an index k for which $IA(k) = i$, $JA(k) = j$. Then $A(i,j) = NA(k)$.

The symbolic factorization also computes a P permutation matrix mentioned in the last section, which may be modified during the numerical factorization process. With this, the numeric factorization may be executed very efficiently. The numeric factorization operation shown in Section 7.1 may be rewritten in this indexed form. The operations in the inner i, j loop of the algorithm are executed only for nonzero terms.

Note, that the sparsity pattern of the resulting factor matrix is different than that of the matrix factored, due to the aforementioned fill-in terms. In fact, the number of nonzero terms of the factor matrix is sometimes orders of magnitude higher. They are still very far from being dense and they are also stored in sparse, indexed form.

While sparse matrix factorization is clearly an advantageous method when considering the outer fringes of the sparse finite element matrices, it could become a burden close to the diagonal. Specifically there are usually fairly dense blocks in finite element matrices as well as completely empty blocks. The following method takes advantage of the best of the sparse and dense approaches.

7.3 Multi-frontal factorization

The premier, industry standard factorization techniques in finite element analysis are in the class of multi-frontal techniques [2]. The simplified idea of such methods is to execute the sparse factorization in terms of dense sub-matrices, the "frontal matrices". The frontal method is really an implementation of the Gaussian elimination by eliminating several rows at the same time.

The multi-frontal method is an extension of this by recognizing that many separate fronts can be eliminated simultaneously by following the elimination pattern of some reordering, like the minimum degree method. Hence, the method takes advantage of the sparsity preserving nature of the reordering and the efficiency provided by the dense block computations.

Let us review the method in terms of our A matrix as follows. Permute and partition A as

$$P_1 A P_1^T = \begin{bmatrix} E_1 & C_1^T \\ C_1 & B_1 \end{bmatrix}.$$

P_1 is the permutation matrix computed by the reordering step, but possibly slightly modified to assure that the inverse of the $s_1 \times s_1$ sub-matrix E_1 exists. The first block elimination or factorization step is then

$$P_1 A P_1^T = \begin{bmatrix} I_1 & 0 \\ C_1 E_1^{-1} & I_{n-1} \end{bmatrix} \begin{bmatrix} E_1 & 0 \\ 0 & B_1 - C_1 E_1^{-1} C_1^T \end{bmatrix} \begin{bmatrix} I_1 & E_1^{-1} C_1^T \\ 0 & I_{n-1} \end{bmatrix}.$$

Here the subscript for I_1 indicates the solution step rather than the size of the identity matrix. The size is s_1 as indicated by the size of E_1. The process is repeated by taking

$$A_2 = B_1 - C_1 E_1^{-1} C_1^T.$$

We attack the next "front" by permuting and partitioning as

$$P_2 A_2 P_2^T = \begin{bmatrix} E_2 & C_2^T \\ C_2 & B_2 \end{bmatrix},$$

and factorizing as above. P_2 is a partition of the P matrix computed by the reordering step and possibly modified to ensure that E_2^{-1} exists. The final factors of

$$\overline{P} A \overline{P}^T = LDL^T$$

are built as

$$
L = \begin{bmatrix}
I_1 & 0 & 0 . 0 \\
C_1 E_1^{-1} & I_2 & 0 . 0 \\
& C_2 E_2^{-1} & I_3 . 0 \\
& & . . \\
& & I_k
\end{bmatrix} ,
$$

and

$$
D = \begin{bmatrix}
E_1 & 0 & 0 & 0 & 0 \\
0 & E_2 & 0 & 0 & 0 \\
0 & 0 & E_3 & 0 & 0 \\
. & . & . & . & . \\
0 & 0 & 0 & 0 & B_k - C_k E_k^{-1} C_k^T
\end{bmatrix} .
$$

D is built from variable size $s_i \times s_i$ diagonal blocks. Note, that the $C_i E_i^{-1}$ sub-matrices are rectangular, extending to the bottom of the factor. The process stops when s_k is desirably small. The $\overline{P}$ is an aggregate of the intermediate P_i permutation matrices. If there were no numerical reasons to modify the original order then $P = \overline{P}$.

It is evident from the form above that the main computational component of the factorization process is the execution of the

$$
B_i - C_i E_i^{-1} C_i^T
$$

step. Since the B_i matrix is an ever shrinking partition of the updated A matrix, this step is called a matrix update. The size of the E_i matrix is the rank of the update. Linear algebra libraries, such as LAPACK [1] have readily available high performance routines to execute such operations. The C_i sub-matrices of course are still sparse, therefore, these operations are executed by indexed operations as shown earlier.

Some words of numerical considerations. In the above we assumed that the inverse of each E_i sub-matrix exist. This is assured by the proper choice of P_i. In fact the quality of the inverse determines the numerical stability of the factorization. If the sparsity pattern-based original permutation matrix P does not produce E_i^{-1}, it is modified (pivoted) to do so.

Finally, this procedure again generalizes to unsymmetric matrices as well.

7.4   Linear system solution

The solution of linear systems is of paramount importance for finite element computations. The direct solution algorithms follow the factorization of the matrices. The reordering executed in the factorization must be observed during the solution phase. Consider the linear system

$$AX = B.$$

Since for the permutation matrices

$$P^T P = I$$

holds, the system may be solved in terms of the factors of the reordered matrix

$$PAP^T = LDL^T$$

as

$$LDL^T PX = PB.$$

There are two solution steps, the forward and the backward substitution. The forward substitution solves for the intermediate results of

$$LY = PB.$$

Here the column permutation of the factorization is executed on the right-hand side B matrix. The backward substitution computes

$$L^T(PX) = D^{-1}Y.$$

The row permutation of the factorization is now observed on the result matrix shown by (PX). Here D^{-1} of course exists as D is comprised of the E_i matrices of the factorization.

For the sake of completeness, these steps for the unsymmetric case are

$$LUQX = PB,$$

with forward substitution of

$$LY = PB,$$

and backward substitution of

$$U(QX) = Y.$$

Naturally, the sparse block structure of the participating matrices is exploited in these computations also.

7.5 Distributed factorization and solution

In industrial practice the finite element matrices are exceedingly large and partitioned with tools like [3] for an efficient solution on multiprocessor computers or network of workstations. The subject of this section is to discuss the computational process enabling such an operation.

Assume that the matrix is partitioned into sub-matrices as follows:

$$
A = \begin{bmatrix}
A_{oo}^1 & & & & & A_{ot}^1 \\
& A_{oo}^2 & & & & A_{ot}^2 \\
& & \cdot & & & \cdot \\
& & & A_{oo}^j & A_{ot}^j & \\
& & & & \cdot & \\
A_{to}^1 & A_{to}^2 & \cdot & A_{to}^j & \cdot & A_{tt}
\end{bmatrix},
$$

where superscript j refers to the j-th partition. The o subscript refers to the interior, while subscript t to the common boundary of the partitions and s is the number of partitions, so $j = 1, 2, \ldots s$. The size of the global matrix is N.

As the solution of a linear system is our goal, let the solution vector and the right-hand side vector be partitioned accordingly.

$$
x = \begin{bmatrix}
x_o^1 \\
x_o^2 \\
\cdot \\
x_o^j \\
\cdot \\
x_t
\end{bmatrix},
$$

and

$$
b = \begin{bmatrix}
b_o^1 \\
b_o^2 \\
\cdot \\
b_o^j \\
\cdot \\
b_t
\end{bmatrix}.
$$

For simplicity of the discussion, we consider only a single vector right-hand side and solution. The j-th processor contains only the j-th partition of the matrix:

$$A^j = \begin{bmatrix} A^j_{oo} & A^j_{ot} \\ A^j_{to} & A^j_{tt} \end{bmatrix},$$

where A^j_{tt} is the complete boundary of the j-th partition that may be shared by several other partitions. Note also, that it is a subset of the global boundary A_{tt}. Similarly the local solution vector component is partitioned

$$x^j = \begin{bmatrix} x^j_o \\ x^j_t \end{bmatrix}.$$

where x^j_t is a partition of x_t as

$$x_t = \begin{bmatrix} x^1_t \\ .. \\ x^j_t \\ .. \\ x^s_t \end{bmatrix}.$$

It is desired that

$$A = LDL^T$$

be computed in partitions. Consider the factor matrices partitioned similarly.

$$L = \begin{bmatrix} L^1_{oo} & & & & \\ & L^2_{oo} & & & \\ & & . & . & \\ & & & L^j_{oo} & \\ & & & & . & . \\ L^1_{to} & L^2_{to} & . & L^j_{to} & . & L_{tt} \end{bmatrix},$$

and

$$D = \begin{bmatrix} D^1_{oo} & & & & \\ & D^2_{oo} & & & \\ & & . & & . \\ & & & D^j_{oo} & \\ & & & & . & . \\ & & . & & . & D_{tt} \end{bmatrix}.$$

Multiplication of the partitioned factors yields

$$A = \begin{bmatrix} L^1_{oo}D^1_{oo}L^{1,T}_{oo} & & & & L^1_{oo}D^1_{oo}L^{1,T}_{to} \\ & L^2_{oo}D^2_{00}L^{2,T}_{oo} & & & L^2_{oo}D^2_{00}L^{2,T}_{to} \\ & & . & & . \\ & & & L^j_{oo}D^j_{00}L^{j,T}_{oo} & L^j_{oo}D^j_{00}L^{j,T}_{to} \\ & & & & . \\ L^1_{to}D^1_{00}L^{1,T}_{oo} & L^2_{to}D^2_{00}L^{2,T}_{oo} & . & L^j_{to}D^j_{00}L^{j,T}_{oo} & . & L_{tt}D_{tt}L^T_{tt} + \Sigma^s_{j=1}L^j_{to}D^j_{oo}L^{j,T}_{to} \end{bmatrix}.$$

The terms of the partitioned factors are obtained in several steps and some of them (L_{tt}, D_{tt}) are not explicitly computed. First, the factorization of the interior of the j-th partition is executed.

$$A^j = \begin{bmatrix} A^j_{oo} & A^j_{ot} \\ A^j_{to} & A^j_{tt} \end{bmatrix} = \begin{bmatrix} L^j_{oo} & 0 \\ L^j_{to} & I \end{bmatrix} \begin{bmatrix} D^j_{oo} & 0 \\ 0 & \overline{A}^j_{tt} \end{bmatrix} \begin{bmatrix} L^{j,T}_{oo} & L^{j,T}_{to} \\ 0 & I \end{bmatrix},$$

where the identity matrices are not computed, they are only presented to make the matrix equation algebraically correct. This step produces the L^j_{oo}, L^j_{to} and the D^j_{oo} local factor components. The $\overline{A}^j_{tt}$ sub-matrix is the local Schur complement of the the j-th partition:

$$\overline{A}^j_{tt} = A^j_{tt} - L^j_{to} D^j_{oo} L^{j,T}_{to},$$

with $A_{tt} = \Sigma^s_{j=1} A^j_{tt}$. Next the individual Schur complement matrices are summed up as

$$\overline{A}_{tt} = \Sigma^s_{j=1} \overline{A}^j_{tt}$$

to create the global Schur complement. Finally, the global Schur complement is factored as:

$$\overline{A}_{tt} = \overline{L}_{tt} \overline{D}_{tt} \overline{L}^T_{tt}.$$

Note the bar over the factor terms as these are the factors of the global Schur complement, not the original boundary partition.

The partitioned solution of

$$LDL^T x = b$$

will also take multiple steps. The forward substitution in the partitioned form is

$$\begin{bmatrix} L^j_{oo} D^j_{oo} & 0 \\ L^j_{to} D^j_{oo} & I \end{bmatrix} \begin{bmatrix} y^j_o \\ y^j_t \end{bmatrix} = \begin{bmatrix} b^j_o \\ b^j_t \end{bmatrix}.$$

Here the I sub-matrix is included only to assure compatibility and

$$b_t = \begin{bmatrix} b^1_t \\ .. \\ b^j_t \\ .. \\ b^s_t \end{bmatrix}.$$

The forward substitution on the first block may be executed for the interior of all partitions first:

$$y^j_o = [L^j_{oo} D^j_{oo}]^{-1} b^j_o.$$

The second block of the forward solve for each partition yields:

$$y_t^j = b_t^j - L_{to}^j D_{oo}^j y_o^j,$$

which then has to be summed up for all partitions as:

$$y_t = \Sigma_{j=1}^s y_t^j.$$

The global boundary solution is a complete forward-backward substitution of

$$\overline{L}_{tt}\overline{D}_{tt}\overline{L}_{tt}^T x_t = y_t.$$

The partitions' interior solutions will be finalized from the backward step of

$$\begin{bmatrix} L_{oo}^{j}{}^T & L_{to}^{j}{}^T \\ 0 & I \end{bmatrix} \begin{bmatrix} x_o^j \\ x_t^j \end{bmatrix} = \begin{bmatrix} y_o^j \\ x_t^j \end{bmatrix}.$$

The first equation yields

$$x_o^j = L_{oo}^{j}{}^{-T}(y_o^j - L_{to}^{j}{}^T x_t^j),$$

where x_o^j is the final result in the interior of the j-th partition.

Note, that the permutation matrix was ignored in the above presentation for simplicity, however, the process works with it also. Computer implementation aspects of the technology are discussed in [8].

7.6 Factorization and solution case studies

The cost of the multi-frontal factorization is of $O(n^2 f_{avg})$, where n is the matrix size and f_{avg} is the average front size. The cost of the solution is $O(n f_{avg} m)$, where the number of vectors (the number of columns in the B matrix) is denoted by m. The range of m depends on the application, but could easily be in the thousands. Nevertheless, the cost of the factorization is the dominant cost of linear system solutions, hence the reason for the first case study example.

The size of matrices in practical problems, even when talking about components, is significant. This is largely due to the fact that the automated mesh generators tend to "over-mesh" physical models. To retain fine local details in one part of the model, other parts which do not need detailed results are nonetheless meshed with small elements. This is done because it takes less effort to create a mesh with a single overall element size than a mesh with a gradation of element sizes. Additionally, models with fine geometric detail - for example, fillets around holes and chamfers around edges - force automatic mesh generators to put small elements in these regions even though these details may be structurally insignificant.

For the studying the factorization issue, we consider a component arising in the automobile industry, a crankshaft casing shown in Figure 7.1.

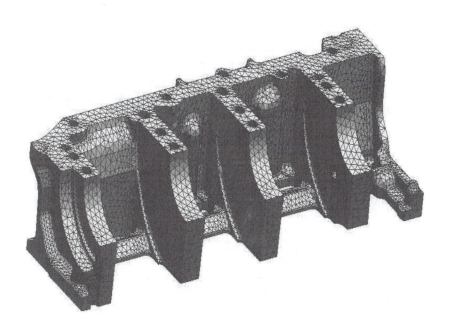

FIGURE 7.1 Crankshaft casing finite element model

Most such models are automatically meshed with solid elements, and in such models the rotational degrees of freedom are all eliminated by the procedure explained in detail in Chapter 5. Therefore, the number of free degrees of freedom (DOF) is roughly the half of the global degrees of freedom in the model minus the boundary conditions. The statistics of the model with a four way partitioning are shown on Table 7.1.

TABLE 7.1

Size statistics of casing component model

Model	Nodes	Elements	g-size	f-size
complete partition	213,470	131,416	1,280,009	638,058
			t-size	f-size
1	55,312	33,937	1,161	164,745
2	50,608	31,424	2,580	154,404
3	53,301	34,351	10,068	169,771
4	50,178	31,704	8,298	158,832

There are some observations to be made. It is interesting to note that the sizes of the partitions are very close, within 10 percent of each other. More importantly, the partition boundary and the front sizes, in specific, tend to follow the partition sizes, the largest partition (3) has the largest boundary and front size, latter shown in Table 7.2.

TABLE 7.2

Computational statistics of casing component model

Model	Front size (max)	Factor size (Kterms)	Factorization time (sec)	Boundary time (sec)
complete partition	4,335	372,132	573.1	–
1	3,729	80,087	112.2	23.8
2	3,447	80,739	120.5	37.8
3	3,816	86,907	135.0	70.5
4	3,585	75,452	108.5	55.2

Table 7.2 also demonstrates the computational complexity of the factorization of this model and the times reported are CPU times. The statistics are from the factorization on a workstation using 4 processor nodes with 1.5 GHz clock speed. As the machine was dedicated, the CPU times are indicative of elapsed performance.

The factor size and factorization time also largely follow the front size order of the partitions. Ultimately, the number of terms in the factor matrix has a direct correlation with the factorization time.

It is also noticeable that the boundary time (Schur complement computation) is significant, it ranges from 20 to almost 70 percent of the partition factorization time. This is related to the t-sizes reported in Table 7.1. The total t-size of the problem was 12,213 as the individual t-sizes largely overlap. Nevertheless, the longest total execution time (205.5 sec for partition 3) is almost one third of the unpartitioned solution time. Computer implementation aspects of the technology are discussed in [8].

We demonstrate the computational complexity of linear system solutions with another case study example of a model from the aerospace industry. The model consisted of approximately 34 million node points and elements. Table 7.3 contains statistics of the matrices in a linear statics analysis.

TABLE 7.3
Linear static analysis matrix statistics

K	number of rows	nonzero terms	max terms
matrix	204 million	3.8 billion	54
F	number of rows	number of columns	max terms
matrix	204 million	1	999
Factor	number of rows	nonzero terms	max front
matrix	35.7 million	51.1 billion	6,995

The direct linear static analysis was executed on a workstation with 8 (1.9 GHz) CPUs. The total linear static solution required 338 minutes of elapsed time. Of this, the factorization alone required approximately 100 minutes and the direct solve about 30 minutes. The amount of I/O processed was 3.04 Terabytes with a 758 Gigabytes disk footprint.

In practical circumstances many different load scenarios are used, resulting in multiple columns of the F matrix. These systems nowadays are also modeled by CAD systems using various solid modeling techniques. This fact leads to somewhat denser matrices, in contrast to the sparse matrices of shell models of a car body or an airplane fuselage.

Direct solution of such systems requires enormous memory and disk resources due to the size of the factor matrices. The advantages of an iterative solution is quite clear and often exploited in industrial practice.

7.7   Iterative solution of linear systems

It is very clear that the cost of the factorization operation is significant and especially so when the model results in dense matrices. There are some computational solutions when the factorization cannot be avoided, however, simple linear system solutions may be more efficiently executed by iterations.

Let us consider the linear statics problem of

$$Ku = F,$$

where the partition designation is omitted for simplicity of the presentation. It is safe to assume that we address the f-partition problem.

The simplest iterative solution of this system may be found by splitting the stiffness matrix into two additive components as

$$K = K_1 - K_2,$$

Then the system may be presented in terms of these components as

$$K_1 u - K_2 u = F,$$

providing a very simple scheme of

$$K_1 u_i = K_2 u_{i-1} + F,$$

where u_i is the i-th iterative solution. When the inverse of the K_1 component exists and any reasonable splitting would be aimed for that, then

$$u_i = K_1^{-1} K_2 u_{i-1} + K_1^{-1} F.$$

The process may be started with $u_0 = 0$ and it is known to converge as long as

$$||K_1^{-1} K_2|| \leq 1.$$

The above is a necessary condition for the convergence of an iterative solution of a system based on a particular splitting.

Naturally the inverse should also be easy to compute and the Jacobi method [5] well known to engineers, is based on simply splitting the diagonal of the stiffness matrix off:

$$K_1 = diag(K(j, j)), j = 1, 2, \ldots, n,$$

a reasonable strategy as long as the stiffness matrix does not contain very small diagonal terms. The iteration scheme presented above simplifies to a termwise formula for every $j = 1, 2, \ldots, n$.

$$u_i(j) = \frac{1}{K(j,j)}(F(j) - \sum_{k=1}^{j-1} K(j,k)u_{i-1}(k)).$$

Above convergence condition adjusted for the Jacobi method is

$$\|K_1^{-1}K_2\| = max_{1 \leq j \leq n} \sum_{k \neq j} |\frac{K(j,k)}{K(j,j)}| < 1,$$

from which it follows that

$$\sum_{k \neq j} |K(j,k)| < |K(j,j)|; j = 1, \ldots, n$$

is required. This translates into the requirement of the diagonal dominance of the K matrix, not at all in odds with finite element stiffness matrices.

The most successful iterative technique in engineering practice today is the conjugate gradient method [4] minimizing the functional

$$G(u) = \frac{1}{2}u^T K u - u^T F,$$

whose first derivative (the gradient) is the residual of the linear system

$$\frac{dG}{du} = Ku - F = -r.$$

The method is a series of approximate solutions of the form

$$u_i = u_{i-1} + \alpha_i p_{i-1},$$

and the consecutive residuals are

$$r_i = r_{i-1} - \alpha_i K p_{i-1}.$$

The method's mathematical foundation is rooted in the Ritz-Galerkin principle that proposes to select such iterative solution vectors u_i for which the residual is orthogonal (hence the conjugate in the name) to a Krylov subspace generated by K and the initial residual r_0.

From the recursive application of the principle emerge the distance coefficients as

$$\alpha_i = \frac{r_{i-1}^T r_{i-1}}{p_{i-1}^T K p_{i-1}}$$

and the search direction of

$$p_i = r_i + \beta_i p_{i-1}.$$

The relative improvement of the solution is measured by

$$\beta_i = \frac{r_i^T r_i}{r_i^T r_i}.$$

The process is initialized as

$$u_0 = 0, r_0 = F, p_0 = r_0.$$

The conjugate gradient method algorithm is using above formulae in a specific order:

For $i = 1, 2, 3, \ldots$ until convergence compute:

Distance: α_i

Approximate solution: u_i

Current residual: r_i

Relative improvement: β_i

New search direction: p_i

If $||r_i|| \le \epsilon$, stop

End loop i.

In above ϵ is a certain threshold. The method nicely generalizes to the unsymmetric case and known as the biconjugate gradient method. Its underlying principle is still the orthogonalization of the successive residuals to the continuously generated Krylov subspace. Due to the unsymmetric nature of the matrix, there are left handed and right handed sequences residuals

$$r_i = r_{i-1} - \alpha_i K p_{i-1},$$

$$s_i = s_{i-1} - \alpha_i K^T q_{i-1}.$$

corresponding to left and right search directions

$$p_i = r_i + \beta_i p_{i-1},$$

$$q_i = s_i + \beta_i q_{i-1}.$$

The coefficients are combining the two sides as

$$\alpha_i = \frac{s_{i-1}^T r_{i-1}}{q_{i-1}^T K p_{i-1}}$$

and

$$\beta_i = \frac{s_i^T r_i}{s_{i-1}^T r_{i-1}}.$$

Ultimately, however, the next approximate solution is of the same form

$$u_i = u_{i-1} + \alpha_i p_{i-1},$$

but the process may only be stopped when both residual norms,

$$||r_i||, ||s_i||,$$

are less than a certain threshold.

7.8 Preconditioned iterative solution technique

There are also variations of iterative solutions when the matrix is preconditioned to accelerate the convergence. The essence of preconditioning is by premultiplying the problem with a suitable preconditioning matrix as

$$P^{-1} K u = P^{-1} F.$$

The conjugate gradient algorithm allows the implicit application of the preconditioner during the iterative process. The preconditioner is commonly presented as an inverse matrix, because of the goal of approximating the inverse of the system matrix. Clearly with the selection of

$$P = K$$

the solution is trivial, since

$$P^{-1} K u = K^{-1} K u = I u = K^{-1} F.$$

The cost of the computation of the preconditioner is of course in this case the cost of the factorization. The iterative solution itself is a forward-backward substitution.

   More practical approaches use various incomplete factorizations of the matrix

$$K \approx CC^T = P.$$

In an incomplete factorization process only those terms of the factor matrix are computed that correspond to nonzero terms of the original matrix. The fill-in terms are omitted. There are several variations of this approach, such as also computing fill-in terms, but only above a certain threshold value. Another extension is to compute the fill-in terms inside a certain band of the matrix.

   It is another commonly used approach to exploit the fact that the matrix is assembled from finite element matrices. In that case the preconditioners are various factorizations of the element matrices as

$$K_e \approx C_e C_e^T = P_e,$$

assembled into a global preconditioner as

$$P = \Sigma_{e=1}^n P_e.$$

Such approaches are described in [9] and [10]. The preconditioning approach is the most successful however when the preconditioner captures and exploits some physics modeling specific information.

   There are cases when the stiffness matrix may consist of two components, one that is easy to solve and one that is difficult. An example is static aero-elastic analysis, where there is a K_s structural stiffness matrix and a matrix K_a representing the aero-elastic effects. The static aero-elastic solution is described by the equation of form

$$(K_s + K_a)u = F.$$

K_s is generally sparse and banded and can be solved directly by itself at a reasonable cost. The K_a matrix is developed by a theory where every point of the structure that is touched by air is coupled to every other point touching air. This leads to a dense, largely un-banded, and unsymmetric matrix.

   In general, an unsymmetric problem is at least twice as expensive to solve as a symmetric problem, and the other unfortunate characteristics of K_a result in an order of magnitude increase in solution cost. Therefore, preconditioning the static aero-elastic equation with

$$P = K_s$$

and reordering to place the aero effects on the right hand side results in an iterative solution scheme:

$$u_i = K_s^{-1}(F - K_a u_{i-1}).$$

The inverse operation is done with a factorization followed by a forward-backward substitution with the vector of unknowns from the right hand side. The structural stiffness matrix acts as a pre-conditioner. u_0 is set to zero, resulting in u_1 being the static solution due to the structural effects only. Only one matrix factorization is required for all the iterations. The troublesome K_a is never factorized.

The physics of the problem dictates that the terms in K_s are much larger than in K_a because the structure is stiffer than the air it intersects. This leads to a rapid convergence of this solution in connection with, for example, the bi-conjugate gradient method. The cost of the coupled aero-elastic solution by this approach is generally no more that twice the cost of a structure only solution.

There are several other classical splitting methods, such as the Gauss-Seidel or the successive over-relaxation methods [13]. The concept in itself is valuable for engineers when they combine different physical phenomena and they know the coupling and partitioning a priori.

There are also other minimization based methods, such as the generalized minimum residual method [12]. There is also another class of solutions based on an adaptive idea [11]. Both of these are subjects of strong academic interest but have not yet been proven as generally useful in the industry as the conjugate gradient method.

References

[1] Anderson, E. et al; LAPACK user's guide, 2nd ed., SIAM, Philadelphia, 1995

[2] Duff, I. S. and Reid, J. K.; The multi-frontal solution of indefinite sparse symmetric linear systems, ACM Trans. Math. Softw. Vol. 9, pp. 302-325, 1983

[3] George, A. and Liu, J. W. H.; The evolution of the minimum degree ordering algorithm, SIAM Review, Vol. 31, pp. 1-19, 1989

[4] Hestenes, M. R.; and Stiefel, E.; Methods of conjugate gradients for solving linear systems, Journal Res. National Bureau of Standards, Vol. 49, pp. 409-436, 1952

[5] Jacobi, C. G. J.; Über eines leichtes Verfahren die in der Theorie der Sëcularstörungen vorkommenden Gleichungen numerisch aufzulösen, J. Reine Angewandte Math., Vol. 30., pp. 51-94, 1846

[6] Liu, J. W. H.; The minimum degree ordering with constraints, SIAM J. of Scientific and Statistical Computing, Vol. 10, pp. 1136-1145, 1988

[7] Karypis, G. and Kumar, V.; ParMETIS:Parallel graph partitioning and sparse matrix library, University of Minnesota, 1998

[8] Mayer, S.; Distributed parallel solution of very large systems of linear equations in the finite element method, PhD Thesis, Technical University of Munich, 1998

[9] Poschmann, P.; and Komzsik, L.; Iterative solution technique for finite element applications, Journal of Finite Element Analysis and Design, Vol. 14, No. 4, pp. 373-381, 1993

[10] Komzsik, L., Sharapov, I., Poschmann, P.; A preconditioning technique for indefinite linear systems, Journal of Finite Element Analysis and Design, Vol. 26, No. 3, pp. 253-258, 1997

[11] Rüde, U.; Fully adaptive multigrid methods, SIAM Journal of Numerical Analysis, Vol. 30, pp. 230-248, 1993

[12] Saad, Y. and Schultz, M. H.; GMRES: a generalized minimum residual algorithm for solving nonsymmetric linear systems, SIAM, Journal of Scientific and Statistical Computing, Vol. 7, pp. 856-869, 1986

[13] Varga, R. S.; Matrix Iterative Analysis, Prentice-Hall, Englewod Cliffs, New Jersey, 1962

8

Static Condensation

The K_{aa} matrix is free of constraints and ready for analysis. However, for computational advantages it may be further reduced. Let us partition the remaining degrees of freedom into two groups. One such physical partitioning may be based on considering the boundary degrees of freedom as one, and the interior degrees of freedom as the other partition. This is a single-level, single-component static condensation, the topic of Section 8.1. One of the first publications related to this topic is [2].

It is also possible to first partition the model into components and apply the boundary reduction for each. This is the single-level, multiple-component condensation of Section 8.2. Finally, the two techniques may be recursively applied yielding the multiple-level, multiple-component method of Section 8.3.

All of these methods allow the possibility of parallel processing. Due to the cost of reduction and back-transformation, however, the computational complexity of solving a certain problem may not change. The static condensation methods produce computationally exact results when applied to the linear static problem. The following sections detail the static condensation technique.

8.1 Single-level, single-component condensation

Let us consider the linear statics problem of

$$K_{aa}u_a = F_a$$

and the simple finite element model shown in Figure 8.1 where we marked the boundary partition with t and the interior with o.

The matrix partitioning corresponding to the model partitioning shown in Figure 8.1 is simply:

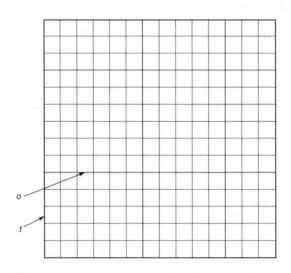

FIGURE 8.1 Single-level, single-component partitioning

$$K_{aa} = \begin{bmatrix} K_{oo} & K_{ot} \\ K_{to} & K_{tt} \end{bmatrix}.$$

The static problem is partitioned accordingly:

$$K_{aa}u_a = \begin{bmatrix} K_{oo} & K_{ot} \\ K_{to} & K_{tt} \end{bmatrix} \begin{bmatrix} u_o \\ u_t \end{bmatrix} = \begin{bmatrix} F_o \\ F_t \end{bmatrix}.$$

Let us introduce a transformation matrix

$$T = \begin{bmatrix} I_{oo} & G_{ot} \\ 0 & I_{tt} \end{bmatrix},$$

where

$$G_{ot} = -K_{oo}^{-1}K_{ot}$$

is the static condensation matrix. Substituting

$$T\overline{u}_a = u_a$$

and pre-multiplying by T^T yields

$$T^T K_{aa} T \overline{u}_a = T^T F_a,$$

or

$$\overline{K}_{aa}\overline{u}_a = \overline{F}_a.$$

The latter equation in details reads

$$\begin{bmatrix} K_{oo} & 0_{ot} \\ 0_{to} & \overline{K}_{tt} \end{bmatrix} \begin{bmatrix} \overline{u}_o \\ u_t \end{bmatrix} = \begin{bmatrix} F_o \\ \overline{F}_t \end{bmatrix}.$$

Here

$$\overline{u}_o = u_o - G_{ot}u_t,$$

and

$$\overline{K}_{tt} = K_{tt} + K_{to}G_{ot}$$

is the Schur complement. Finally the modified load is

$$\overline{F}_t = F_t + G_{ot}^T F_0.$$

This results in the reduced (statically condensed) problem of

$$\overline{K}_{tt}u_t = \overline{F}_t$$

from which the reduced, boundary solution is computed. To obtain the interior solution, one solves

$$K_{oo}\overline{u}_o = F_o$$

followed by the back-transformation

$$u_o = \overline{u}_o + G_{ot}u_t.$$

The importance of the order of operations in the efficiency of numerical algorithms is critical and we often trade formulation for efficiency. A case in point is the method of reducing matrices via the Schur complement. The above matrix formulation is conceptually simple and easier to implement, however, it is not very efficient for large scale analysis.

In practice the G_{ot} transformation is not calculated explicitly. Instead, the K_{aa} matrix is partially factored as shown in the last chapter; only degrees of freedom in the o partition are eliminated:

$$\begin{bmatrix} L_{oo} & 0_{ot} \\ L_{to} & I_{tt} \end{bmatrix} \begin{bmatrix} D_{oo} & 0_{ot} \\ 0_{to} & \overline{K}_{tt} \end{bmatrix} \begin{bmatrix} L_{oo}^T & L_{to}^T \\ 0_{to} & I_{tt} \end{bmatrix} \begin{bmatrix} u_o \\ u_t \end{bmatrix} = \begin{bmatrix} F_o \\ F_t \end{bmatrix}.$$

The partial factor matrices can also be used to derive the partitioned solution. This indicates that they will produce identical results (within computational errors) and are merely a different order of operations, not a change in the method of solution. Here now the Schur complement is of the form

$$\overline{K}_{tt} = K_{tt} - L_{to}D_{oo}L_{to}^T.$$

Developing the second row of the above matrix equation one gets

$$L_{to}D_{oo}L_{oo}^T u_o + \overline{K}_{tt}u_t + L_{to}D_{oo}L_{to}^T u_t = F_t.$$

Expanding the first row yields

$$L_{oo}D_{oo}L_{oo}^T u_o + L_{oo}D_{oo}L_{to}^T u_t = F_o.$$

Executing a forward substitution on the interior we get an intermediate interior solution as

$$L_{oo}^T u_o + L_{to}^T u_t = (L_{oo}D_{oo})^{-1}F_o = \overline{u}_o.$$

Note that this intermediate solution is different from that computed in the matrix form. Similarly, the modified boundary load in terms of the partial factors is different and computed as

$$\overline{F}_t = F_t - L_{to}D_{oo}\overline{u}_o.$$

Despite these intermediate step differences, the final solution is the same. Premultiplying this by $L_{to}D_{oo}$ and subtracting it from the developed form of the second row, we get the reduced problem of

$$\overline{K}_{tt}u_t = \overline{F}_t.$$

The reduced matrix factorization of

$$\overline{K}_{tt} = L_{tt}D_{tt}L_{tt}^T$$

produces the reduced solution

$$L_{tt}D_{tt}L_{tt}^T u_t = \overline{F}_t$$

using forward-backward substitution. The interior solution is finally computed via the earlier calculated partial factors

$$u_o = L_{oo}^{-T}(\overline{u}_o - L_{to}^T u_t).$$

This process, when using sparse matrix factorization techniques described in Chapter 7 that recognize the sparsity pattern of finite element matrices, is very efficient for large problems.

8.2   Computational example

To demonstrate the static condensation principle, let us consider the following small numerical example.

$$K_{aa} = \begin{bmatrix} 2 & 1 & 0 \\ 1 & 3 & 1 \\ 0 & 1 & 4 \end{bmatrix}.$$

To execute the static condensation with the single component case, partition the problem into

$$K_{oo} = \begin{bmatrix} 2 & 1 \\ 1 & 3 \end{bmatrix} \quad K_{ot} = \begin{bmatrix} 0 \\ 1 \end{bmatrix}$$

$$K_{to} = \begin{bmatrix} 0 & 1 \end{bmatrix} \quad K_{tt} = \begin{bmatrix} 4 \end{bmatrix}.$$

The static condensation matrix is calculated as

$$G_{ot} = -\begin{bmatrix} 2 & 1 \\ 1 & 3 \end{bmatrix}^{-1} \begin{bmatrix} 0 \\ 1 \end{bmatrix} = -\begin{bmatrix} 3/5 & -1/5 \\ -1/5 & 2/5 \end{bmatrix} \begin{bmatrix} 0 \\ 1 \end{bmatrix} = \begin{bmatrix} 1/5 \\ -2/5 \end{bmatrix}.$$

The static condensation transformation matrix becomes

$$T = \begin{bmatrix} 1 & 0 & 1/5 \\ 0 & 1 & -2/5 \\ 0 & 0 & 1 \end{bmatrix}.$$

The Schur complement result of the static condensation is

$$\overline{K}_{tt} = \begin{bmatrix} 4 \end{bmatrix} + \begin{bmatrix} 0 & 1 \end{bmatrix} \begin{bmatrix} 1/5 \\ -2/5 \end{bmatrix} = \begin{bmatrix} 18/5 \end{bmatrix}.$$

Finally, the statically condensed matrix is

$$\overline{K} = \begin{bmatrix} 2 & 1 & 0 \\ 1 & 3 & 0 \\ 0 & 0 & 18/5 \end{bmatrix}.$$

The procedure is rather straightforward, but in practice the transformation and condensation matrices are not built explicitly, as was shown in the last section and will be demonstrated in the following.

As these operations are instrumental in the following chapters, we continue the example with the solution of a system with the same matrix

$$\overline{K}u = \begin{bmatrix} 2 & 1 & 0 \\ 1 & 3 & 1 \\ 0 & 1 & 4 \end{bmatrix} \begin{bmatrix} 3 \\ 2 \\ 1 \end{bmatrix} = \begin{bmatrix} 8 \\ 10 \\ 6 \end{bmatrix} = F.$$

The first, matrix based solution scheme is as follows. The modified load vector boundary component with the condensation matrix computed above is

$$\overline{F}_t = \begin{bmatrix} 6 \end{bmatrix} + \begin{bmatrix} 1/5 & -2/5 \end{bmatrix} \begin{bmatrix} 8 \\ 10 \end{bmatrix} = \begin{bmatrix} 18/5 \end{bmatrix}.$$

The boundary solution, using the Schur complement from the condensation part of the example above is

$$u_t = K_{tt}^{-1}\overline{F}_t = [18/5]^{-1}[18/5] = [1].$$

This is already a component of the final solution and agrees with the analytic value. The intermediate interior solution is

$$\overline{u}_o = K_{oo}^{-1}F_o = \begin{bmatrix} 2 & 1 \\ 1 & 3 \end{bmatrix}^{-1} \begin{bmatrix} 8 \\ 10 \end{bmatrix} = \begin{bmatrix} 14/5 \\ 12/5 \end{bmatrix}.$$

Finally, the actual interior solution is

$$u_o = \overline{u}_o + G_{ot}u_t = \begin{bmatrix} 14/5 \\ 12/5 \end{bmatrix} + \begin{bmatrix} 1/5 \\ -2/5 \end{bmatrix}[1] = \begin{bmatrix} 3 \\ 2 \end{bmatrix}.$$

This also agrees with the analytic solution.

The partial factor based computational solution scheme is as follows. The partial factorization of the matrix results in:

$$K = \begin{bmatrix} 1 & 0 & 0 \\ 1/2 & 1 & 0 \\ 0 & 2/5 & 1 \end{bmatrix} \begin{bmatrix} 2 & 0 & 0 \\ 0 & 5/2 & 0 \\ 0 & 0 & 18/5 \end{bmatrix} \begin{bmatrix} 1 & 1/2 & 0 \\ 0 & 1 & 2/5 \\ 0 & 0 & 1 \end{bmatrix}.$$

The resulting components of the partial factorization are

$$L_{oo} = \begin{bmatrix} 1 & 0 \\ 1/2 & 1 \end{bmatrix} \quad L_{to} = [0 \ 2/5]$$

$$D_{oo} = \begin{bmatrix} 2 & 0 \\ 0 & 5/2 \end{bmatrix} \quad K_{tt} = [18/5].$$

Note that the 18/5 term in above, while it is identical to the explicitly computed Schur complement, was a side result of the partial factorization due to the update of that process and as such, it was free.

The intermediate interior solution in terms of the partial factors is obtained from the forward only substitution of the equation

$$L_{oo}D_{oo}\overline{u}_o = F_o$$

or

$$\begin{bmatrix} 1 & 0 \\ 1/2 & 1 \end{bmatrix} \begin{bmatrix} 2 & 0 \\ 0 & 5/2 \end{bmatrix} \begin{bmatrix} \overline{u}_o(1) \\ \overline{u}_o(2) \end{bmatrix} = \begin{bmatrix} 8 \\ 10 \end{bmatrix}$$

as

$$\overline{u}_o = \begin{bmatrix} 4 \\ 12/5 \end{bmatrix}.$$

Note that the intermediate solution is different then the matrix formulation. The modified load vector in terms of the partial factors and the intermediate interior result is

$$\overline{F} = F_t - L_{to}D_{oo}\overline{u}_o,$$

numerically

$$[6] - [0 \; 2/5] \begin{bmatrix} 2 & 0 \\ 0 & 5/2 \end{bmatrix} \begin{bmatrix} 4 \\ 12/5 \end{bmatrix} = [18/5.] .$$

This results in the boundary solution of

$$u_t = \begin{bmatrix} 1 \end{bmatrix} ,$$

as before and as required by the analytic result. The final interior results are computed from the backward only substitution of

$$L_{oo}^T u_0 = \overline{u}_o - L_{to}^T u_t ,$$

or

$$\begin{bmatrix} 1 & 1/2 \\ 0 & 1 \end{bmatrix} u_o = \begin{bmatrix} 4 \\ 12/5 \end{bmatrix} - \begin{bmatrix} 0 \\ 2/5 \end{bmatrix} [1] ,$$

resulting in the correct values of

$$u_o = \begin{bmatrix} 3 \\ 2 \end{bmatrix} .$$

The complete solution is assembled as

$$u = \begin{bmatrix} 3 \\ 2 \\ 1 \end{bmatrix} .$$

The computational solution clearly required more steps to execute, however, at a smaller overall computational cost mainly due to the avoidance of explicit matrix forming and inverse computations.

8.3 Single-level, multiple-component condensation

In this case, the model is first subdivided into multiple components and the static condensation is executed simultaneously on all components. The partitioning is executed automatically by applying specialized graph partitioning techniques to the finite element model. It is important that the partitioning produce close to equal partitions and a minimal boundary between the partitions, a tough problem indeed.

A well-known method of separating the graph of the matrix into partitions is the nested dissection [4]. The class of multilevel partitioning methods of [1] and [3] are also widely accepted in the industry.

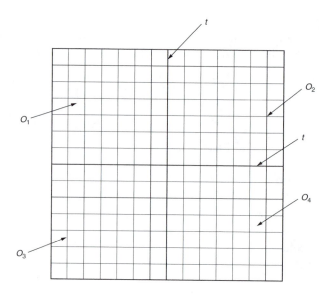

FIGURE 8.2 Single-level, multiple-component partitioning

Let us consider the model partitioning shown in Figure 8.2. Let t_{12} denote the common boundary between the geometric partitions o_1 and o_2, in essence the upper half of the vertical divider line on the figure. Similarly t_{13} is the common boundary between o_1 and o_3 and so on. Finally, the t_0 will be the boundary that is shared by all partitions, in this case the central vertex of the vertical and horizontal dividers.

Ordering the interior partitions o_i first, followed by the boundary partitions t_{ij} and finished by the shared boundary t_0 as

$$o_1 \; o_2 \; o_3 \; o_4 \; t_{12} \; t_{13} \; t_{24} \; t_{34} \; t_0 \, ,$$

will result in the following K_{aa} stiffness matrix pattern:

$$
\begin{bmatrix}
K_{oo}^1 & & & & & & K_{o_1 t_{12}} & K_{o_1 t_{13}} & & & & K_{o_1 t_0} \\
& K_{oo}^2 & & & & & K_{o_2 t_{12}} & & K_{o_2 t_{24}} & & & K_{o_2 t_0} \\
& & K_{oo}^3 & & & & & K_{o_3 t_{13}} & & K_{o_3 t_{34}} & K_{o_3 t_0} \\
& & & K_{oo}^4 & & & & & K_{o_4 t_{24}} & K_{o_4 t_{34}} & K_{o_4 t_0} \\
K_{t_{12} o_1} & K_{t_{12} o_2} & & & K_{tt}^{12} & K_{tt}^{12,13} & K_{tt}^{12,24} & & & & K_{tt}^{12,0} \\
K_{t_{13} o_1} & & K_{t_{13} o_3} & & K_{tt}^{13,12} & K_{tt}^{13} & & K_{tt}^{13,34} & K_{tt}^{13,0} \\
& K_{t_{24} o_2} & & K_{t_{24} o_4} & K_{tt}^{24,12} & & K_{tt}^{24} & K_{tt}^{24,34} & K_{tt}^{24,0} \\
& & K_{t_{34} o_3} & K_{t_{34} o_4} & & K_{tt}^{34,13} & K_{tt}^{34,24} & K_{tt}^{34} & K_{tt}^{34,0} \\
K_{t_0 o_1} & K_{t_0 o_2} & K_{t_0 o_3} & K_{t_0 o_4} & K_{tt}^{0,12} & K_{tt}^{0,13} & K_{tt}^{0,24} & K_{tt}^{0,34} & K_{tt}^{0}
\end{bmatrix}.
$$

Here $K_{oo}^i, i = 1 \ldots 4$ contains the stiffness matrix partition corresponding to the interior of the i-th partition. $K_{o_i t_{ij}}$ designates the i-th partition's boundary-coupling with the j-th, $j = 1 \ldots 4$. The terms K_{tt}^{ij} contain the stiffness matrix partition of the common boundary between the i-th and j-th partitions. The terms $K_{tt}^{ij,kl}$ denote coupling between the ij-th and kl-th boundaries, $k, l = 1 \ldots 4$. Finally, $K_{tt}^{ij,0}$ is the coupling between the ij-th and the common boundary.

The specific sparsity pattern of the boundary partitions is reflective of the particular partitioning and ordering. In general cases there may be a rather difficult boundary and boundary-coupling sparsity pattern, therefore, a combined K_{tt} term of

$$
K_{tt} =
\begin{bmatrix}
K_{tt}^{12} & K_{tt}^{12,13} & K_{tt}^{12,24} & & K_{tt}^{12,0} \\
K_{tt}^{13,12} & K_{tt}^{13} & & K_{tt}^{13,34} & K_{tt}^{13,0} \\
K_{tt}^{24,12} & & K_{tt}^{24} & K_{tt}^{24,34} & K_{tt}^{24,0} \\
& K_{tt}^{34,13} & K_{tt}^{34,24} & K_{tt}^{34} & K_{tt}^{34,0} \\
K_{tt}^{0,12} & K_{tt}^{0,13} & K_{tt}^{0,24} & K_{tt}^{0,34} & K_{tt}^{0}
\end{bmatrix}
$$

is introduced, where the total boundary is comprised of all the local boundaries:

$$t = t_{12} + t_{13} + t_{24} + t_{34} + t_0.$$

Let us also introduce

$$K_{ot}^1 = \begin{bmatrix} K_{o_1 t_{12}} & K_{o_1 t_{13}} & 0 & 0 & K_{o_1 t_0} \end{bmatrix},$$

$$K_{ot}^2 = \begin{bmatrix} K_{o_2 t_{12}} & 0 & K_{o_2 t_{24}} & 0 & K_{o_2 t_0} \end{bmatrix},$$

$$K_{ot}^3 = \begin{bmatrix} 0 & K_{o_3 t_{13}} & 0 & K_{o_3 t_{34}} & K_{o_3 t_0} \end{bmatrix},$$

and

$$K_{ot}^4 = \begin{bmatrix} 0 & 0 & K_{o_4 t_{24}} & K_{o_4 t_{34}} & K_{o_4 t_o} \end{bmatrix}.$$

With the above, the matrix partitioning used for the following discussions is

$$K_{aa} = \begin{bmatrix} K_{oo}^1 & & & & K_{ot}^1 \\ & K_{oo}^2 & & & K_{ot}^2 \\ & & K_{oo}^3 & & K_{ot}^3 \\ & & & K_{oo}^4 & K_{ot}^4 \\ K_{to}^1 & K_{to}^2 & K_{to}^3 & K_{to}^4 & K_{tt} \end{bmatrix}.$$

The static problem in the general multiple (n) component case following the latter notation is:

$$K_{aa} u_a = \begin{bmatrix} K_{oo}^1 & & & K_{ot}^1 \\ & \ddots & & \\ & & K_{oo}^i & K_{ot}^i \\ & & & \ddots \\ K_{to}^1 & \cdot & K_{to}^i & \cdot & K_{tt} \end{bmatrix} \begin{bmatrix} u_o^1 \\ \cdot \\ u_o^i \\ \cdot \\ u_t \end{bmatrix} = \begin{bmatrix} F_o^1 \\ \cdot \\ F_o^i \\ \cdot \\ F_t \end{bmatrix} = F_a.$$

The condensation matrix of the i-th component, following the notation of the preceding section and introducing a superscript for the component, is

$$G_{ot}^i = -K_{oo}^{-1,i} K_{ot}^i.$$

The multiple component transformation matrix is

$$T = \begin{bmatrix} I_{oo}^1 & & & G_{ot}^1 \\ & \ddots & & \cdot \\ & & I_{oo}^i & G_{ot}^i \\ & & & \ddots & \cdot \\ & & & & I_{tt} \end{bmatrix}.$$

Using the pre-multiplication by T^T and the substitution of $T\bar{u}_a = u_a$ as in the single component case results in

$$T^T K_{aa} T \bar{u}_a = T^T F_a,$$

or

$$\overline{K}_{aa} \bar{u}_a = \overline{F}_a.$$

In detail this multiple component partitioned form of the condensed problem is

$$\begin{bmatrix} K_{oo}^1 & & & \\ & \ddots & & \\ & & K_{oo}^i & \\ & & & \ddots & \\ & & & & \overline{K}_{tt} \end{bmatrix} \begin{bmatrix} \bar{u}_o^1 \\ \cdot \\ \bar{u}_o^i \\ \cdot \\ u_t \end{bmatrix} = \begin{bmatrix} F_o^1 \\ \cdot \\ F_o^i \\ \cdot \\ \overline{F}_t \end{bmatrix}.$$

The multiple component Schur complement is of the form

$$\overline{K}_{tt} = K_{tt} + \Sigma_{i=1}^n K_{ot}^{T,i} G_{ot}^i.$$

The modified interior solution components formally are

$$\overline{u}_o^i = u_0^i - G_{ot}^i u_t,$$

and the modified boundary load is

$$\overline{F}_t = F_t + \Sigma_{i=1}^n G_{ot}^{T,i} F_o^i.$$

The efficient computational technology relies on the partial factors, as shown earlier. The modified boundary load in terms of the partial factors of the components is

$$\overline{F}_t = F_t - \Sigma_{i=1}^n L_{to}^i D_{oo}^i \overline{u}_o^i,$$

where the intermediate interior solutions are

$$\overline{u}_o^i = (L_{oo}^i D_{oo}^i)^{-1} F_o^i.$$

In terms of the partial factors of the components the reduced matrix is

$$\overline{K}_{tt} = K_{tt} - \Sigma_{i=1}^n L_{to}^i D_{oo}^i L_{to}^{T,i},$$

and while small, it is very dense. The reduced solution again is of the form

$$\overline{K}_{tt} u_t = \overline{F}_t.$$

The final interior solution of the components is calculated from

$$u_o^i = L_{oo}^{-T,i} (\overline{u}_o^i - L_{to}^{T,i} u_t).$$

Since the computations related to the interior of the components are independent of each other, this method is naturally applicable to parallel computers. However, care must be applied when dealing with the t partition as multiple components contribute to it. The problem becomes even more complex on a parallel computer; special synchronization logic is needed to deal with the contributions to the t partition. This topic is beyond our focus.

The single-level, multiple-component method has a notable computational shortcoming because the size of the t partition increases proportionally to the number of components. To overcome this problem, a multiple-level static condensation may also be used.

8.4   Multiple-level static condensation

Let us reconsider the simple finite element problem again with a different partitioning, as shown in Figure 8.3. In the figure the nodes in set t^{12} represent the common boundary between partitions o_1 and o_2. Similarly set t^{34} is the boundary between partitions o_3, o_4. These boundaries represent the first level of partitioning. The second level partitioning is represented by set t^{1234} which is the collection of the boundary sets.

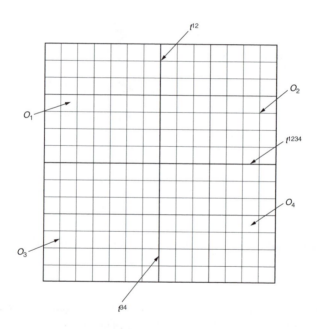

FIGURE 8.3 Multiple-level, multiple-component partitioning

The stiffness matrix structure corresponding to this partitioning is discussed in the following. It is important to point out that the coupling matrices from an interior domain to the various boundaries are different. They are noted by a specific superscript structure. For example the $K_{ot}^{1,12}$ coupling is from the 1st interior domain to the common boundary between the 1st and the 2nd components. For simplicity the coupling from the ith domain to the final boundary is noted as $K_{ot}^{i,0}$. Note, that only the upper triangle of the symmetric matrix is shown.

$$K_{aa} = \begin{bmatrix} K_{oo}^1 & K_{ot}^{1,12} & & & & & K_{ot}^{1,0} \\ & K_{oo}^2 & K_{ot}^{2,12} & & & & K_{ot}^{2,0} \\ & & K_{tt}^{12} & & & & K_{tt}^{12,0} \\ & & & K_{oo}^3 & K_{ot}^{3,34} & K_{ot}^{3,0} \\ & sym & & K_{oo}^4 & K_{ot}^{4,34} & K_{ot}^{4,0} \\ & & & & K_{tt}^{34} & K_{tt}^{34,0} \\ & & & & & K_{tt}^0 \end{bmatrix}.$$

The component condensation matrices for this arrangement are

$$G_{ot}^{1,12} = -K_{oo}^{-1,1} K_{ot}^{1,12},$$

$$G_{ot}^{2,12} = -K_{oo}^{-1,2} K_{ot}^{2,12},$$

$$G_{ot}^{3,34} = -K_{oo}^{-1,3} K_{ot}^{3,34},$$

and

$$G_{ot}^{4,34} = -K_{oo}^{-1,4} K_{ot}^{4,34}.$$

Note, that the $^{-1,i}$ superscript marks the inverse of the interior of the i-th component.

Two transformation matrices may be built to condense the components to the boundaries between them. They are

$$T_{12} = \begin{bmatrix} I_{oo}^1 & & G_{ot}^{1,12} \\ & I_{oo}^2 & G_{ot}^{2,12} \\ & & I_{tt}^{12} \\ & & & I_{oo}^3 \\ & & & & I_{oo}^4 \\ & & & & & I_{tt}^{34} \\ & & & & & & I_{tt}^0 \end{bmatrix}$$

and

$$T_{34} = \begin{bmatrix} I_{oo}^1 \\ & I_{oo}^2 \\ & & I_{tt}^{12} \\ & & & I_{oo}^3 & & G_{ot}^{3,34} \\ & & & & I_{oo}^4 & G_{ot}^{4,34} \\ & & & & & I_{tt}^{34} \\ & & & & & & I_{tt}^0 \end{bmatrix}.$$

The identity matrices' sub- and superscripts mark their sizes and locations. The successive application to the stiffness matrix accomplishes the condensation of the two pairs of components to their respective common boundaries, completing the first level of the reduction.

The second (the highest in our case) level of reduction requires the elimination of the boundary-coupling terms of the last column via a new set of condensation matrices. The following are still simple boundary condensation matrices from the interior of the components to the final boundary:

$$G_{ot}^{1,0} = -K_{oo}^{-1,1} K_{ot}^{1,0},$$

$$G_{ot}^{2,0} = -K_{oo}^{-1,2} K_{ot}^{2,0},$$

$$G_{ot}^{3,0} = -K_{oo}^{-1,3} K_{ot}^{3,0},$$

and

$$G_{ot}^{4,0} = -K_{oo}^{-1,4} K_{ot}^{4,0}.$$

Another two condensation matrices are needed to eliminate the boundary-coupling terms. They are:

$$G_{tt}^{12,0} = -\overline{K}_{tt}^{-1,12} K_{tt}^{12,0}$$

and

$$G_{tt}^{34,0} = -\overline{K}_{tt}^{-1,34} K_{tt}^{34,0}.$$

Note, that these condensation matrices are calculated with the $\overline{K}_{tt}$ terms, that are the already statically condensed lower level boundary components (local Schur complements). The transformation matrix for the second level reduction is formed in terms of previous condensation matrices.

$$T_0 = \begin{bmatrix} I_{oo}^1 & & & & & & G_{ot}^{1,0} \\ & I_{oo}^2 & & & & & G_{ot}^{2,0} \\ & & I_{tt}^{12} & & & & G_{tt}^{12,0} \\ & & & I_{oo}^3 & & & G_{ot}^{3,0} \\ & & & & I_{oo}^4 & & G_{ot}^{4,0} \\ & & & & & I_{tt}^{34} & G_{tt}^{34,0} \\ & & & & & & I_{tt}^0 \end{bmatrix}.$$

The transformation matrix executing the multiple-component multiple-level static condensation for our case is

$$T = T_{12} T_{34} T_0.$$

Note, that more than two transformation matrices may be in the first level and also more than two levels may be applied. Applying the transformation matrix as

$$\overline{K}_{aa} = T^T K_{aa} T$$

results in the following stiffness matrix structure

$$\overline{K}_{aa} = \begin{bmatrix} K_{oo}^1 & & & & & & \\ & K_{oo}^2 & & & & & \\ & & \overline{K}_{tt}^{12} & & & & \\ & & & K_{oo}^3 & & & \\ & & & & K_{oo}^4 & & \\ & & & & & \overline{K}_{tt}^{34} & \\ & & & & & & \overline{K}_{tt}^0 \end{bmatrix}.$$

Note, that this reduction step is not marked by a different partition name but with $\overline{K}_{aa}$, as the size of this matrix has not been reduced. On the other hand the density has been radically reduced. The terms denoted by $\overline{*}$ are the condensed (Schur complement) terms. The actual execution of the solution of the condensed static problem is now straightforward.

The following chart illustrates the effect of the static condensation to the stiffness matrix, independently of which version was used. Of course the $\overline{K}_{tt}$ is not a direct partition of K_{aa}.

$$\begin{bmatrix} K_{aa} & \\ & [\overline{K}_{tt}] \end{bmatrix}.$$

Note, that the static condensation process is computationally exact, meaning that apart from the errors introduced by the floating point arithmetic, no other approximation error occurred.

It is also important to mention that the multiple-component static condensation technique enables very efficient solution of large linear statics problems when executed on parallel computers or network of workstations. The automatic partitioning technique mentioned in the beginning of this chapter should also take into consideration the cost of decomposition of the domains and the possible equivalency of those costs. Specifically, the number of nonzero terms in the rows of the component matrices (front size) should also be considered during domain decomposition.

8.5 Static condensation case study

To demonstrate the practicalities of the static condensation, let us consider the industrial example of the crankshaft of an automobile. An example of such a structural component is shown in Figure 8.4.

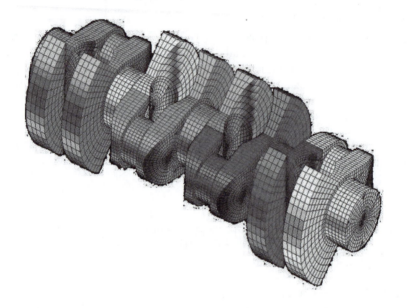

FIGURE 8.4 Automobile crankshaft industrial example

The model had 550,132 node points and 369,468 4-noded tetrahedral elements. An eight-component static condensation resulted in the characteristics shown in Table 8.1. The internal nodes are the o^i partitions. The boundary nodes are the t^i partitions and their union produces the t partition.

The original model was partitioned automatically and the number of elements and the number of degrees of freedom in each component demonstrate the quality of the partitioning.

The effect of the multiple-component static condensation for a linear static analysis of the model is shown in Table 8.2. The serial execution is the original model without static condensation and the parallel is the static condensation version on eight processors. The computer was the same as described in

TABLE 8.1
Component statistics of crankshaft model

Internal nodes	Boundary nodes	Elements	DOF
76,265	503	50,775	230,304
73,167	1,339	49,912	223,518
59,133	1,899	39,671	183,096
67,245	1,035	46,155	204,840
66,898	3,179	45,990	210,231
77,805	2,111	53,185	239,748
63,504	2,528	43,943	198,096
58,999	2,438	39,837	183,111

Chapter 7, but it is immaterial here as the relation between the serial and the parallel executions is of importance here.

TABLE 8.2
Performance statistics of crankshaft model

	Elapsed min:sec	CPU seconds	I/O GBytes	Memory Mwords
Serial	71:14	3,982.	36.67	120
Parallel	24:46	1,442.	11.08	50

The I/O reported is not the total disk storage usage, it is the amount of data transferred during the analysis. As such, it is much larger than the actual disk requirement, due to repeated data access.

It is noticeable that the memory and I/O requirements (the main advantage of using the condensation) are significantly reduced. This effect is given even when the analysis is executed on a serial machine. On the other hand, when the condensed version is running on multiple processors, there is an additional performance advantage in execution times.

The speedup is not very good. This is due to the fact that the data is that of a complete analysis job, the factorization and solution steps are only a part of that. This fact is limiting the achievable speedup.

References

[1] Barnard, S. T. and Simon, H. D.; Fast multilevel implementation of recursive spectral bisection for partitioning unstructured problems, Concurrency: Practice and Experience, Vol. 6(2), pp. 101-117, Wiley and Sons, New York, 1994

[2] Guyan, R. J.; Reduction of stiffness and mass matrices, AIAA Journal, Vol. 3, pp. 380-390, 1965

[3] Karypis, L. and Kumar, V.; A fast and high quality multilevel scheme for partitioning irregular graphs, Tech. Report TR 95-035, Department of Computer Science, University of Minnesota, Minneapolis, 1995

[4] Liu, J. W. H.; The role of elimination trees in sparse factorization, SIAM J. of Matrix Analysis and Applications, Vol. 11, pp. 134-172, 1990

9

Real Spectral Computations

In practice large scale eigenvalue problems are solved with specific techniques discussed in this chapter. These techniques also provide the foundation for the dynamic reduction methods of this second part of the book. These computations are usually executed by employing a spectral transformation followed by a robust eigenvalue solution technique, such as the Lanczos method.

9.1   Spectral transformation

In industrial applications the eigenvalue spectrum of interest is very wide and non-homogeneous. Regions with closely spaced eigenvalues are followed by empty segments and vice versa. The convergence of most commonly used eigenvalue methods is slow in that case.

The motivation of the spectral transformation is to modify the spectral distribution to find the eigenvalues more efficiently. This is done in the form of a transformed eigenvalue:

$$\mu = \frac{1}{\lambda - \lambda_s},$$

where λ_s is an appropriately chosen eigenvalue shift. The graphical representation of this is a hyperbola shifted to the right by λ_s in the (μ, λ) coordinate system. As shown in Figure 9.1, this transformation enables one to find closely spaced eigenvalues in the neighborhood of λ_s in a well separated form on the μ axis.

Applying the spectral transformation to the algebraic eigenvalue problem of

$$A\underline{x} = \lambda\underline{x}$$

in the form of

$$\lambda = \frac{1}{\mu} + \lambda_s,$$

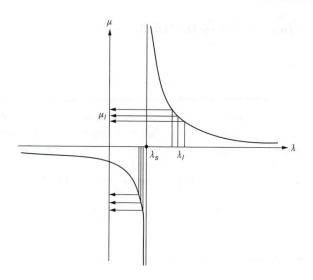

FIGURE 9.1 Spectral transformation

we get

$$(A - \lambda_s I)^{-1} \underline{x} = \mu \underline{x},$$

or

$$\overline{A}\underline{x} = \mu \underline{x}.$$

In practical computations the $\overline{A}$ matrix of course is never explicitly formed. Any time when the eigenvalue method requires an operator multiplication of

$$z = \overline{A}\underline{x},$$

one executes the

$$(A - \lambda_s I) = LDL^T$$

symmetric factorization followed by the

$$LDL^T z = \underline{x}$$

forward-backward substitution. In the iterations only the substitution steps are executed until a new spectral transformation is done. While the cost of the substitution operations may be seemingly too high in the place of the matrix-vector multiplication, the numerical advantages gained clearly outweigh the costs. The explicit inverse of $(A - \lambda_s I)$ is generally a full matrix, while its factor L, albeit denser than A, is still a sparse matrix in most applications. The spectral transformation step may be executed at multiple $\lambda_{s1}, \lambda_{s2}, \ldots$ locations, enabling the traversal of very wide frequency ranges of interest that are commonplace in the industry.

The possible problem of λ_s being identical (or too close) to an eigenvalue must be resolved by a singularity detection mechanism. The singularity detection is done in the factorization operation similarly to the technique described in Section 5.2. It is based on monitoring the terms of the D diagonal factor matrix relative to the corresponding original terms in $A - \lambda_s I$. If their ratio is too high (λ_s is too close to an eigenvalue), the value of λ_s is perturbed and the factorization is repeated.

Another advantage of the spectral transformation is that the factorization at the various shifts produces a mechanism to monitor the distribution of eigenvalues. Namely, the number of negative terms on the D factor matrix is equivalent to the Sturm number. The Sturm number is the number of alterations of sign in the Sturm sequence $d_0, d_1, \ldots d_{n-1}$, where $d_i = det(A_i - \lambda_s I)$ is the determinant of the leading i-th principal sub-matrix of the shifted matrix with $d_0 = 1$.

It follows that the Sturm number also indicates the number of eigenvalues located to the left of the current shift λ_s in the spectrum. This is a tool exploited by all industrial eigenvalue solution software packages to verify that all eigenvalues in a region bounded by two λ_s values are found. If the number of eigenvalues found in a region is less then the differences in the Sturm numbers at the boundaries of the region, the region is bisected and additional iterations are executed as needed. If the number of eigenvalues found in the region exceeds the Sturm count difference, then an error has occurred and one or more spurious eigenvalues were found.

9.2   Lanczos reduction

The most wide spread and robust method for the solution of industrial eigenvalue problems is the Lanczos method [6]. Let us consider the canonical eigenvalue problem

$$A\underline{x} = \lambda\underline{x},$$

with a real, symmetric A. Here $\underline{x}$ are the eigenvectors of the original problem; the underlining is used to distinguish it from the soon to be introduced x Lanczos vectors.

The Lanczos method generates a set of orthogonal vectors X_n such that:

$$X_n^T X_n = I,$$

where I is the identity matrix of order n and

$$X_n^T A X_n = T_n,$$

where A is the original real, symmetric matrix and T_n is a tridiagonal matrix of form

$$T_n = \begin{bmatrix} \alpha_1 & \beta_1 & & & \\ \beta_1 & \alpha_2 & \beta_2 & & \\ & \cdot & \cdot & \cdot & \\ & & \beta_{n-2} & \alpha_{n-1} & \beta_{n-1} \\ & & & \beta_{n-1} & \alpha_n \end{bmatrix}.$$

By multiplication we get the following equation:

$$A X_n = X_n T_n,$$

By equating columns on both sides of the equation we get:

$$A x_k = \beta_{k-1} x_{k-1} + \alpha_k x_k + \beta_k x_{k+1},$$

where $k = 1, 2, ..n - 1$ and x_k are the k-th columns of X_n. For any $k < n$ the following is also true:

$$A X_k = X_k T_k + \beta_k x_{k+1} e_k^T,$$

where e_k is the k-th unit vector containing unit value in row k and zeroes elsewhere. Its presence is only needed to make the matrix addition operation compatible. This equation will be very important in the error-bound calculation. The following starting assumption is made:

$$\beta_0 x_0 = 0.$$

By reordering we obtain the following Lanczos recurrence formula:

$$\beta_k x_{k+1} = A x_k - \alpha_k x_k - \beta_{k-1} x_{k-1}.$$

The coefficients β_k and α_k are defined as

$$\beta_k = \sqrt{|x_{k+1}^T x_{k+1}|},$$

and

$$\alpha_k = x_k^T A x_k.$$

The process is continued. Sometimes, due to round-off error in the computations, the orthogonality between the Lanczos vectors is lost. This is remedied by executing a Gram-Schmidt orthogonalization step at certain k values:

$$\gamma_i = x_{k+1} x_i \quad i = 1, .., k,$$

and

$$x_{k+1} = x_{k+1} - \Sigma_{i=1}^k \gamma_i x_i.$$

This can be a very time-consuming step and in the industry the orthogonalization is only executed against a certain selected set of i indices, not all.

The solution of the eigenvalue problem is now based on the tridiagonal matrix T_n as follows:

$$T_n u_i = \lambda_i u_i.$$

The eigenvalues are invariant under the reduction and the eigenvectors are recovered as

$$\underline{x} = X_n u_i.$$

In practice the Lanczos reduction is executed only up to a certain number of steps, say $j << n$. The approximated residual error in the original solution following a partial Lanczos reduction [7] can be calculated as:

$$||r_j|| = ||A\underline{x} - \lambda_i \underline{x}|| = ||AX_j u_i - \lambda_i X_j u_i|| = ||(AX_j - \lambda_i X_j)u_i||,$$

where $i = 1, 2, \ldots j$, and j is the number of Lanczos steps executed. Furthermore,

$$||r_j|| = ||(AX_j - X_j T_j)u_i|| = ||(\beta_j x_{j+1} e_j^T)u_i|| = \beta_j ||e_j^T u_i||,$$

assuming the norm of the Lanczos vector x_{j+1} is unity. All above norms are Euclidean norms. Taking advantage of the structure of the unit vector we can simplify into the following scalar form:

$$||r_j|| = \beta_j |u_{ji}|,$$

where u_{ji} is the j-th (last) term in the u_i eigenvector.

The last equation gives a convergence monitoring tool. When the error norm is less than the required tolerance ϵ and the value of j is higher than the number of eigenvalues required, the reduction process can be stopped. The

beauty of this convergence criterion is that only the eigenvector of the tridiagonal problem has to be found, which is inexpensive compared to finding the
eigenvector of the physical (size n) problem.

9.3   Generalized eigenvalue problem

In engineering finite element analysis the eigenvalue problem appears in connection with at least two matrices. This is the focus of our interest here, the
generalized linear eigenvalue problem of

$$K\phi - \lambda M\phi = 0,$$

with the assumption that the matrices are symmetric and the partition subscript of the matrices is omitted for simplicity. Since the frequency range of
interest in these problems is at the lower end of the spectrum, it is advisable
to use the spectral transformation introduced in Section 9.1:

$$\mu = \frac{1}{\lambda - \lambda_s}.$$

This will change the problem into

$$(K - \lambda_s M)\phi = \frac{1}{\mu}M\phi,$$

which results in a canonical form amenable to the Lanczos algorithm

$$\mu\phi = (K - \lambda_s M)^{-1}M\phi.$$

Despite its appearance, the matrix operator will be symmetric if the M matrix is used in the inner product of the Lanczos iteration. The process scheme,
consisting of the spectral transformation, the block tridiagonal reduction and
the eigenvalue solution steps is depicted in Figure 9.2.

The eigenvectors are invariant under the spectral transformation and the
eigenvalues may be recovered as

$$\lambda = \frac{1}{\mu} + \lambda_s.$$

The industrial standard block Lanczos method [5] carries out the Lanczos
recurrence with several vectors, called a block, simultaneously. A step of the
Lanczos recurrence algorithm using blocks of vectors, formulated for the generalized eigenvalue problem is

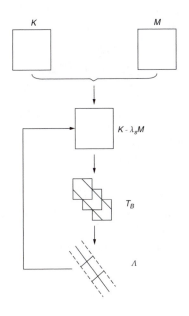

FIGURE 9.2 Generalized solution scheme

$$R_{k+1} = (K - \lambda_s M)^{-1} M Q_k - Q_k A_k - Q_{k-1} B_k^T,$$

where

$$A_k = Q_k^T M (K - \lambda_s M)^{-1} M Q_k,$$

and

$$R_{k+1} = Q_{k+1} B_{k+1}.$$

Q_{k+1} is an n by b matrix with M-orthonormal columns (the Lanczos vectors) and B_{k+1} is a b by b upper triangular matrix, n being the problem size and b the block size, obtainable by the QR decomposition [2].

The block orthogonalization may be formulated as

$$Q_{k+1} = Q_{k+1} - \Sigma_{i=1}^k Q_i \Gamma_i,$$

where

$$\Gamma_i^T = Q_{k+1}^T M Q_i \quad i = 1, .., k$$

and the vectors are mass orthogonalized.

The initial values are $Q_0 = 0$ and R_0 is a collection of b pseudo-random vectors at the start. This process, executed j times, results in a block tridiagonal matrix of the form

$$
T_B = \begin{bmatrix} A_1 & B_2^T & & & \\ B_2 & A_2 & B_3^T & & \\ & \cdot & \cdot & \cdot & B_j \\ & & & B_j & A_j \end{bmatrix}.
$$

With appropriately chosen Givens transformations this tridiagonal matrix is reduced into a scalar tridiagonal matrix T_J. If b is the number of vectors in a block, then the size of T_J is $J = jb$, $(J << n)$. The solution of the eigenvalue problem of

$$
T_J \psi = \mu \psi,
$$

will be addressed in the next section. The eigenvalues of the original large problem are invariant under the transformations resulting in the tridiagonal form. Finally, to find the eigenvectors of the original problem a back-transformation of form

$$
\phi_a = Q_J \psi
$$

is required. The Q_J matrix is a collection of the Lanczos vector blocks:

$$
Q_J = \begin{bmatrix} Q_1 & Q_2 & .. & Q_j \end{bmatrix}.
$$

This process then may be repeated until all eigenvalues (and corresponding eigenvectors) of the required frequency range are found.

9.4   Eigensolution computation

The Lanczos process produced the T_J reduced tridiagonal matrix. Consider the

$$
T_J \psi_i = \mu_i \psi_i
$$

eigenvalues problem. The i in the above equations is the index of the eigenvalue of the tridiagonal matrix, $i = 1, 2, \ldots J$.

Since the eigenvalues are invariant under the transformation to tridiagonal form, the μ_i eigenvalues of the tridiagonal matrix, the Ritz values, are approximations of the λ eigenvalues of the original problem apart from the spectral transformation.

The approximations to the eigenvectors of the original problem (Ritz vectors) are calculated from the eigenvectors of the tridiagonal problem via the multiplication by the Lanczos vectors shown earlier.

To solve the eigenvalue problem the QR iteration [2] may be used, among other well known methods, such as the QL method or the method of bisection in the symmetric case. The QR method is based on a decomposition of the T_J matrix into the form

$$T_J - \omega I = Q^1 R^1,$$

where R^1 is an upper triangular matrix and Q^1 (there are not the Lanczos vectors) contains orthogonal columns:

$$Q^{1,T} Q^1 = I.$$

The presence of ω accounts for a diagonal shift to aid the stability of this decomposition. The process is followed by iterating as follows:

$$T_J^1 = R^1 Q^1 + \omega I.$$

Pre- and post-multiplying gets

$$T_J^2 = Q^{1,T} T_J^1 Q^1.$$

This is a congruence transformation, therefore, the newly created matrix preserves the eigenvalue spectrum of the old one. By repeatedly applying this procedure, T_J^m is finally of diagonal form, whose elements are the eigenvalues of the original tridiagonal matrix, that is

$$T_J^m = diag(\mu_i),$$

where m is the number of iterations, hopefully, but not necessarily, much less than J. The computation in QR iteration takes advantage of the tridiagonal nature of T_J. It creates orthogonal transformation matrices Q^i containing only the four terms of the Givens rotations [3] that reduce the sub-diagonal terms of R^i to zero. This operation will not be detailed more here.

For the eigenvectors an inverse power iteration procedure [8] will be used. The eigenvectors corresponding to the i-th eigenvalue of the T_J tridiagonal matrix may be determined by the following factorization:

$$T_J - \mu_i I = L_i U_i,$$

where L_i is unit lower triangular and U_i is upper triangular. Gaussian elimination with partial pivoting is used, i.e., the pivotal row at each stage is selected to be the equation with the largest coefficient of the variable being eliminated. Since the original matrix is tridiagonal, at each stage there are only two equations containing that variable. Approximate eigenvectors of the i-th eigenvalue λ_i will be calculated by the simple (since U_i also has only 3 co-diagonals) iterative procedure:

$$U_i\psi_i^{k+1} = \psi_i^k,$$

where ψ_i^0 is random and k is a counter. Practice shows that the convergence of this procedure is so rapid that k only goes to 2 or 3.

This original method also has some significant extensions to deal with special cases such as multiple eigenvalues, which are ignored here. These two steps complete the solution of the eigenvalue problem.

The approximated residual error of the i-th eigenpair is

$$||r^i|| = ||K_{aa}\phi_a^i - \lambda_i M_{aa}\phi_a^i||,$$

where

$$\lambda_i = \frac{1}{\mu_i} + \lambda_s.$$

Based on the discourse in the last section, the error can be estimated by

$$||r_J^i|| = \beta_J|\psi_i(J)|$$

where $\psi_i(J)$ is the J-th (last) term in the ψ_i eigenvector and β_J is the last off-diagonal term of the T_J matrix.

The last equation gives the convergence monitoring tool mentioned earlier. When this error norm is less than a pre-defined tolerance ϵ and the value of J is higher than the number of eigenvalues required, the process can be stopped.

9.5   Distributed eigenvalue computation

Due to the vast sizes of industrial problems a distributed computational scheme also has significant merit here. This section introduces an implicit distributed formulation of the Lanczos method. The formulation relies heavily on the distributed factorization and solution technique introduced in Section 7.4. Let us use the standard Lanczos recurrence for the generalized problem as:

$$\overline{x}_{k+1} = Ax_k - \alpha_k x_k - \beta_{k-1}x_{k-1}, \quad , k = 1.2, \ldots$$

with orthogonality parameter

$$\alpha_k = x_k^T M_{aa} A x_k.$$

The canonical (dynamic) matrix is

$$A = (K_{aa} - \lambda_s M_{aa})^{-1} M_{aa},$$

where the fixed λ_s value represents the spectral transformation. The K_{aa}, M_{aa} matrices are the structural stiffness and mass matrices. The normalization parameter is

$$\beta_k = \sqrt{|\bar{x}_{k+1}^T M_{aa} \bar{x}_{k+1}|},$$

and the next Lanczos vector of the recurrence is computed as

$$x_{k+1} = \bar{x}_{k+1}/\beta_k.$$

The following six steps constitute a distributed execution of this process.

I. Partitioned eigenproblem

The global dynamic matrix is partitioned into sub-matrices as follows:

$$A = \begin{bmatrix} A_{oo}^1 & & & & A_{oa}^1 \\ & A_{oo}^2 & & & A_{oa}^2 \\ & & \cdot & & \cdot \\ & & & A_{oo}^j & A_{oa}^j \\ & & & & \cdot \\ A_{ao}^1 & A_{ao}^2 & \cdot & A_{ao}^j & \cdot & A_{aa} \end{bmatrix},$$

where superscript j refers to the j-th partition, subscript a to the common boundary of the partitions and s is the number of partitions, so $j = 1, 2, \ldots s$, just like in the static condensation case. The size of the global matrix as well as the global eigenvectors is N. A key feature of the distributed algorithm is the partitioning of the global Lanczos vectors accordingly:

$$x = \begin{bmatrix} x_o^1 \\ x_o^2 \\ \cdot \\ x_o^j \\ \cdot \\ x_a \end{bmatrix}.$$

In the distributed execution the j-th processor contains only the j-th partition of:

$$A^j = \begin{bmatrix} A_{oo}^j & A_{oa}^j \\ A_{ao}^j & A_{aa}^j \end{bmatrix},$$

where A^j_{aa} is the complete boundary of the j-th partition that may be shared by several other partitions. Similarly a local Lanczos vector component is built as

$$x^j = \begin{bmatrix} x^j_o \\ x^j_a \end{bmatrix},$$

where x^j_a is a partition of x_a.

II. Partitioned matrix factorization

Since the A matrix is available only in the partitions shown above, the factorization is comprised of several components as shown in Section 7.4. The partitioned factorization of the j-th partition is as follows:

$$A^j = \begin{bmatrix} A^j_{oo} & A^j_{oa} \\ A^j_{ao} & A^j_{aa} \end{bmatrix} = \begin{bmatrix} L^j_{oo} & 0 \\ L^j_{ao} & I \end{bmatrix} \begin{bmatrix} D^j_{oo} & 0 \\ 0 & \overline{A}^j_{aa} \end{bmatrix} \begin{bmatrix} L^{T,j}_{oo} & L^{T,j}_{ao} \\ 0 & I \end{bmatrix},$$

where

$$\overline{A}^j_{aa} = A^j_{aa} - L^j_{ao} D^j_{oo} L^{T,j}_{ao}$$

with $A_{aa} = \Sigma^s_{j=1} A^j_{aa}$. The individual Schur complement matrices of each partition are summed up as

$$\overline{A}_{aa} = \Sigma^s_{j=1} \overline{A}^j_{aa}$$

to create the global Schur complement which is factored:

$$\overline{A}_{aa} = \overline{L}_{aa} \overline{D}_{aa} \overline{L}^T_{aa}.$$

This is the foundation of the distributed Lanczos step.

III. Partitioned substitution

The forward substitution in the partitioned form is

$$\begin{bmatrix} L^j_{oo} D^j_{oo} & 0 \\ L^j_{ao} D^j_{oo} & I \end{bmatrix} \begin{bmatrix} \overline{z}^j_o \\ \overline{z}^j_a \end{bmatrix} = \begin{bmatrix} x^j_o \\ x_a \end{bmatrix}.$$

Here the I sub-matrix is included only to assure compatibility. The forward substitution yields

$$\overline{z}^j_o = [L^j_{oo} D^j_{oo}]^{-1} x^j_o,$$

and

$$\overline{z}^j_a = x_a - L^j_{ao} D^j_{oo} \overline{z}^j_o.$$

This is the boundary component of the new local Lanczos vector. The latter has to be summed up for all partitions as:

$$\bar{z}_a = \Sigma_{j=1}^s \bar{z}_a^j.$$

The global boundary solution is therefore

$$\bar{L}_{aa}\bar{D}_{aa}\bar{L}_{aa}^T z_a = \bar{z}_a.$$

The backward substitution for the partitions

$$\begin{bmatrix} L_{oo}^j{}^T & L_{ao}^j{}^T \\ 0 & I \end{bmatrix} \begin{bmatrix} z_o^j \\ z_a \end{bmatrix} = \begin{bmatrix} \bar{z}_o^j \\ z_a \end{bmatrix},$$

produces

$$z_o^j = L_{oo}^j{}^{-T} (\bar{z}_o^j - L_{ao}^j{}^T z_a),$$

which is the interior component of the local Lanczos vector.

IV. Local matrix multiply

In order to compute the orthogonality parameter, the local Lanczos vector component must be multiplied by the mass matrix as follows

$$y_k^j = \begin{bmatrix} y_o^j \\ y_a^j \end{bmatrix} = \begin{bmatrix} M_{oo}^j & M_{oa}^j \\ M_{ao}^j & M_{aa}^j \end{bmatrix} \begin{bmatrix} x_o^j \\ x_a^j \end{bmatrix} = M^j x_k^j.$$

Note, that the x_a^j, y_a^j local boundary partitions must be updated across all the processes.

V. Implicit distributed Lanczos step

At this point we have calculated both components z_o^j, z_a^j of the local z_k^j vector as well as the y_o^j, y_a^j components of the local y_k^j vector partitions. The appropriate partitions of the last two Lanczos vectors x_k^j, x_{k-1}^j are also available. Hence, the implicit distributed Lanczos recurrence is executed with the following steps.

1. Calculate partition local inner product:

$$\alpha_k^j = y_k^{T,j} z_k^j$$

2. Accumulate global inner product:

$$\alpha_k = \Sigma_{j=1}^s \alpha_k^j$$

3. Execute local Lanczos step:

$$\bar{x}^j_{k+1} = z^j_k - \alpha_k x^j_k - \beta_{k-1} x^j_{k-1}$$

4. Calculate local normalization parameter

$$y^j_{k+1} = M^j \bar{x}^j_{k+1}$$

$$\beta^j_k = (\bar{x}^{T,j}_{k+1} y^j_{k+1})^{1/2}$$

5. Collect global value

$$\beta_k = \sqrt{\Sigma^s_{j=1} (\beta^j_k)^2}$$

6. Produce the next normalized Lanczos vector local component

$$x^j_{k+1} = x^j_{k+1} / \beta_k$$

VI. Distributed orthogonalization scheme

A time-consuming operation in the Lanczos reduction is the orthogonalization. This is done by first calculating the local coefficients

$$\omega^j_i = x^{T,j}_{k+1} M^j x^j_i,$$

and summing up the global values as

$$\omega_i = \Sigma^s_{j=1} \omega^j_i,$$

where $i = 1, 2, .., k - 1$ is the index of the already computed Lanczos vectors. The local orthogonalization is done by

$$x^j_{o,k+1} = x^j_{o,k+1} - \omega_i x^j_{o,i},$$

and

$$x^j_{a,k+1} = x^j_{a,k+1} - \omega_i x^j_{a,i}.$$

This concludes the distributed computational form of the Lanczos reduction.

9.6   Dense eigenvalue analysis

Here we discuss the solution of the free vibration problem when the matrices are dense. Such may occur when the structure is a solid model, such as the automobile crankshaft example shown in the last chapter. This may also occur when dealing with a boundary component of a complex structure; this will be discussed in more detail in the next chapter. Finally, in the coupled fluid-structure application of Chapter 6, the fluid component also produces dense matrices.

While the overall concept of solving these eigenvalue problems is similar to the solution of the global problem in the sense that a reduction is followed by eigenvalue iterations, the applied reduction techniques are different and discussed in the following.

In order to apply the reduction techniques to these problems, the problem is explicitly converted to a single matrix canonical form. Since the origin of the dense matrix may be from different sources, in the following we omit the partition notation when we address the real, symmetric, undamped problem of

$$(K - \lambda M)\phi = 0.$$

Let us assume first that the Cholesky factorization of the dense mass matrix exists

$$M = CC^T.$$

The algorithm for such a factorization of an order n symmetric, positive definite A matrix is a mild variation of the factorization introduced in Chapter 7 and may be written as:

$$C(1,1) = \sqrt{A(1,1)}$$

For $j = 2, n$

$$C(j,1) = A(j,1)/C(1,1)$$

End loop j
For $i = 2, n - 1$

$$C(i,i) = (A(i,i) - \Sigma_{k=1}^{i-1} C(i,k)^2)^{1/2}$$

For $j = i + 1, n$

$$C(j,i) = (A(j,i) - \Sigma_{k=1}^{j-1} C(i,k)C(j,k))/C(i,i)$$

End loop j
End loop i

$$C(n,n) = (A(n,n) - \Sigma_{k=1}^{j-1} C(n,k)^2)^{1/2}$$

With this factorization the eigenvalue problem is converted as

$$C^{-1}K\phi_h - \lambda C^{-1}CC^T\phi_h = 0.$$

Introducing

$$\psi = C^T\phi$$

and substituting yields the canonical problem of

$$A\psi = \lambda\psi.$$

Here

$$A = C^{-1}KC^{-1,T}.$$

It is of course not always possible to factor M as shown above. In this case, we assume that the factorization of a linear combination of the mass and stiffness matrices exist. If that does not hold, a massless mechanism exists in the problem and should be removed as shown in Chapter 5.

Assuming a positive shift value (λ_s) for the linear combination, let us factor

$$K + \lambda_s M = CC^T.$$

A heuristic, but industrially proven selection of such a value is in the form of

$$\lambda_s = \frac{1}{\sqrt{n}}\Sigma_{i=1}^n \frac{K(i,i)}{M(i,i)},$$

where n is the size of the matrices. The following rearrangement is the basis of the canonical form in this case:

$$(K + \lambda_s M - (\lambda + \lambda_s)M)\phi = 0.$$

Pre-multiplying this equation by

$$\frac{-1}{\lambda + \lambda_s}$$

and substituting the above factorization yields

$$(M - \frac{CC^T}{\lambda + \lambda_s})\phi = 0.$$

Introducing

$$\phi_h = C^{-1,T}\psi$$

and pre-multiplying with C^{-1} yields the following canonical form

$$(A - \bar{\lambda}I)\psi = 0.$$

Here

$$A = C^{-1}MC^{-1,T},$$

and

$$\bar{\lambda} = \frac{1}{\lambda + \lambda_s}.$$

The solution of this canonical problem is addressed in the next section.

9.7 Householder reduction technique

The most practical methods for dense matrix reduction are based on the Householder reflection. The Householder [4] reflection matrices are of form

$$P_k = I - 2\frac{v_k v_k^T}{v_k^T v_k}.$$

Consider a dense vector of order n

$$x = \begin{bmatrix} x_1 \\ \dots \\ x_k \\ \dots \\ x_n \end{bmatrix}.$$

Let us select the elements of the v_k vector as follows:

- zeroes in the first $k - 1$ terms
- the value of $x_k + \alpha\ sign(x_k)$ for the kth element, and
- the elements of x vector in the $k + 1$ to the nth element.

Here

$$\alpha = \Sigma_{i=k}^n x_i^2.$$

With such a vector

$$v_k = \begin{bmatrix} 0 \\ \cdot\cdot \\ 0 \\ x_k + \alpha\ sign(x_k) \\ x_{k+1} \\ \dots \\ x_n \end{bmatrix}$$

the following is true

$$P_k x = \begin{bmatrix} x_1 \\ \dots \\ x_{k-1} \\ -\alpha \ sign(x_k) \\ 0 \\ \dots \\ 0 \end{bmatrix}.$$

The index k for both v and P indicates the pivoting location. The transformed (reflected vector) has zeroes in the locations from the $k+1$th to the nth. The kth term is modified, and the first $k-1$ terms are unmodified. The geometric meaning of the operation is that the x vector is reflected through the hyper-plane with normal vector v_k, hence the name.

The application of the Householder process to dense matrices results in a reduction to a compact tridiagonal form. This is executed by recursively multiplying a matrix with transformation matrices based on the Householder transformation. With such steps certain components of some columns and rows of the matrix may be zeroed out.

Let us consider a symmetric A matrix and execute the following sequence of matrix transformations

$$A_r = P_r A_{r-1} P_r$$

where the P_r matrix is a Householder matrix of form:

$$P_r = I - \beta v_r v_r^T$$

with $\beta = 2/(v_r^T v_r)$. We choose the elements of v_r such that the rows $r+1$ to n of a vector may be zeroed out. For example, after the rth step of the process, the first r columns of the matrix contain zeroes below the sub-diagonal and the first r rows will have zeroes beyond the super-diagonal term.

$$A_r = \begin{bmatrix} x & x & 0 & 0 & 0 & 0 \\ x & x & x & 0 & 0 & 0 \\ 0 & x & . & . & . & . \\ 0 & 0 & . & x & x & x \\ 0 & 0 & . & x & x & x \\ 0 & 0 & . & x & x & x \end{bmatrix}.$$

Here the x indicates nonzero terms. It is important to execute these operations in such a manner as to maintain symmetry. In detail the transformation at step r is

$$A_r = (I - \beta_r v_r v_r^T) A_{r-1} (I - \beta_r v_r v_r^T).$$

Introducing

$$p_r = \beta_r A_{r-1} v_r,$$

the transformation is

$$A_r = A_{r-1} - v_r p_r^T - p_r v_r^T + \beta_r v_r (v_r^T p_r) v_r^T.$$

Furthermore with

$$q_r = p_r - \frac{1}{2} v_r \beta_r (v_r^T p_r)$$

the symmetric step is

$$A_r = A_{r-1} - v_r q_r^T - q_r v_r^T.$$

The latter form is practical for computer implementation. Repeated application from $r = 1$ to $n - 2$ will result in

$$A_{n-2} = T$$

where T is tridiagonal.

The importance of this reduction is that it is again a congruence transformation, and as such does not change the eigenvalues of the original matrix. Hence, the eigensolution calculation method for tridiagonal matrices introduced in Section 9.4 is also applicable here.

9.8 Normal modes analysis case studies

To demonstrate the power of the computational methods introduced in this chapter, we solve another practical engineering problem introduced in Chapter 3, the free, undamped vibration of structures. This is called normal modes analysis in the industry. This solution will be the foundation of the dynamic reduction technique of the next chapter. It is also very important in the dynamic analysis of global structures to avoid vibration conflicts with the environment in which the structure is operating. The computational problem is

$$(K - \lambda M)u = 0.$$

The goal of normal modes analysis is to find the natural frequencies (λ) and corresponding vibration shapes (u) of the structure. Mostly the lowest natural frequencies of the structure are of particular interest to the engineer.

FIGURE 9.3 Trimmed car body model

Occasionally the natural frequencies in a certain range are needed, to assure the avoidance of resonance catastrophes or annoying vibrations in the audible range.

Let us consider a trimmed car body automobile model shown in Figure 9.3. Such models have all major components of the car, such as wheels, shocks, windows, etc., incorporated and lead to large sparse eigenvalue problems solved in the industry mostly by the Lanczos method.

TABLE 9.1
Model statistics of trimmed car body

Model data	Number of nodes	Number of shells	Number of solids	Number of rigids
	380,007	361,249	3,762	9,056
Sizes	g	n	f	a
	2,280,042	2,223,139	2,223,109	1,937,282

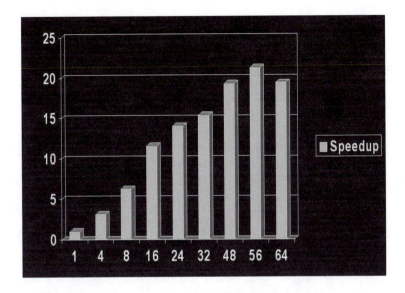

FIGURE 9.4 Speedup of parallel normal modes analysis

The statistics of Table 9.1 show the characteristics of such a finite element model. Note, that they are not the statistics of the illustration model. As it was shown earlier such models usually contain a variety of elements and a large amount of constraints and rigid components.

The task of finding the natural frequencies and corresponding mode shapes of such a model is truly an enormous one. The distributed parallel computation of Section 9.5 enables the feasible execution of this task.

The statistics of the computation encompassing about 900 modes up to 200 Hz are shown on Table 9.2. The analysis was executed on a cluster of 8 workstations, each containing 8 processors with 1.5 GHz clock cycle. The cluster had a 1 Gigabit Ethernet connection.

The elapsed time utilizing 32 processors is already a feasible execution, considering the work environment and time schedule in typical automobile companies. The efficiency above that decreases, but the speedup is still increasing. It peaks at 56 processors, as it is shown in Figure 9.4 where the horizontal axis is the number of processors. More efficient implementation of the computational technique may overcome this limit. Furthermore, the technique may scale to over 100 processors with a larger problem or wider frequency range.

TABLE 9.2
Distributed normal modes analysis statistics

I/O GByte	Elapsed min:sec	Elapsed speedup	Number of processors
2,028.4	523:58	1.00	1
266.9	83:41	6.26	8
191.1	45:16	11.57	16
98.3	34:07	15.35	32
77.8	27:14	19.23	48
67.1	24:41	21.22	56
61.4	27:00	19.40	64

Another model is used to demonstrate the computational complexity of dense component models, also from the automobile industry. A complete engine model, such as shown for example in Figure 9.5 consisted of approximately 12 million node points and 7.5 million elements.

Table 9.3 contains statistics of the matrices in the eigenvalue analysis. The max terms column indicates the maximum number of nonzero terms in the densest column of the matrices. The zero columns of the M matrix is the zero subspace of M. The max front is the maximum front size of the factor matrix.

TABLE 9.3
Normal modes analysis dense matrix statistics

K matrix	number of rows 35.7 million	nonzero terms 1.38 billion	max terms 18,9571
M matrix	number of rows 35.7 million	zero columns 5,317,732	max terms 11
Factor matrix	number of rows 35.7 million	nonzero terms 43.8 billion	max front 30,310

The task of finding the natural frequencies and corresponding mode shapes was executed up to 200 Hz on a workstation with 8 (1.95 GHz) CPUs. The

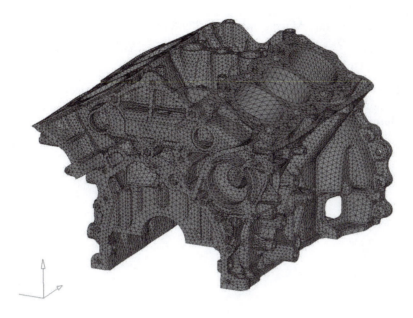

FIGURE 9.5 Engine block model

computation required about 680 minutes elapsed and 100,000 seconds of CPU time using all 8 processors of the workstation in a shared memory fashion. 11.5 Terabytes of I/O was executed and 650 Gigabytes of disk footprint was required. The computational complexity of such industrial normal modes applications is overwhelming.

References

[1] Cullum, J. K. and Willoughby, R. A.; Lanczos algorithms for large symmetric eigenvalue computations, Birkhauser, Boston, 1985

[2] Francis, J. G. F.; The QR transformation I. and II., The Computer Journal, Vol. 4, pp. 265-271, Vol. 5, pp. 332-345, 1961, 1962

[3] Givens, W.; Numerical computation of the characteristic values of a real symmetric matrix, Report ORNL-1574, Oak Ridge National Laboratory,

1954

[4] Householder A. S.; Unitary triangularization of a non-symmetric matrix, J. Assoc. Comp. Mach., Vol 5. pp. 339-342, 1958

[5] Komzsik, L.; The Lanczos method: Evolution and Application, SIAM, Philadelphia, 2003

[6] Lanczos, C.; An iteration method for the solution of eigenvalue problem of linear differential and integral operators, Journal of the National Bureau of Standards, Vol. 49, pp. 409-436, 1952

[7] Parlett, B. N.; The symmetric eigenvalue problem, Prentice-Hall, 1980

[8] Wilkinson, J. H.; The calculation of eigenvectors of co-diagonal matrices, The Computer Journal, Vol. 1, pp 90-92, 1958

10

Complex Spectral Computations

The damped vibration of structures is described by

$$M\ddot{v} + B\dot{v} + Kv = 0,$$

where B is the damping matrix, and $\dot{v}$ refers to the velocity. Executing a Fourier transformation yields the following quadratic eigenvalue problem

$$(M\lambda^2 + B\lambda + K)\phi = 0.$$

The matrices of this quadratic eigenvalue problem may be complex and the problem may also have a left-handed solution

$$\psi^H(M\lambda^2 + B\lambda + K) = 0$$

that is different from the right-hand solution. Here and in the following, H denotes complex conjugate transpose. The solution of this problem usually results in complex eigenvalues. In order to solve the quadratic eigenvalue problem, a transformation is executed to convert the original quadratic problem to a linear problem of twice the size.

10.1 Complex spectral transformation

First the problem is rewritten as a 2 by 2 block linear problem:

$$\lambda \begin{bmatrix} M & 0 \\ 0 & I \end{bmatrix} \begin{bmatrix} \dot{\phi} \\ \phi \end{bmatrix} + \begin{bmatrix} B & K \\ -I & 0 \end{bmatrix} \begin{bmatrix} \dot{\phi} \\ \phi \end{bmatrix} = 0,$$

where $\dot{\phi} = \lambda\phi$. This equation is now linear, albeit, has some serious limitations. For example, one would need to invert both the mass and the damping matrices and an unsymmetric, indefinite matrix built from the damping and the stiffness matrices, in order to reach a solution. Even though the explicit inverses are not needed, the numerical decompositions on either of these matrices may not be very well defined.

A way to improve on this is to execute a complex spectral transformation, similar to the one introduced in Section 9.1 as

$$\mu = \frac{1}{\lambda - \lambda_0},$$

where now λ_0 is a possibly complex shift. By substituting and reordering we get:

$$\mu \begin{bmatrix} \dot{\phi} \\ \phi \end{bmatrix} = \begin{bmatrix} -B - M\lambda_0 & -K \\ I & -\lambda_0 I \end{bmatrix}^{-1} \begin{bmatrix} M & 0 \\ 0 & I \end{bmatrix} \begin{bmatrix} \dot{\phi} \\ \phi \end{bmatrix}.$$

The latter equation is a canonical form of

$$\mu x = Ax,$$

and the corresponding left-handed problem of

$$\mu y^H = y^H A,$$

where

$$A = \begin{bmatrix} -B - M\lambda_0 & -K \\ I & -\lambda_0 I \end{bmatrix}^{-1} \begin{bmatrix} M & 0 \\ 0 & I \end{bmatrix}.$$

The form allows the singularity of the participating matrices, however, their zero subspaces may not coincide. This is a much lighter and more practical restriction.

10.2   Biorthogonal Lanczos reduction

The industry preferred solution for such problems is the bi-orthogonal Lanczos method. In practical implementations the A matrix is not built explicitly; the matrix and its inverse are implicitly used in the eigensolution as shown in the next section.

The block implementation of the bi-orthogonal Lanczos method [1] generates two sets of bi-orthonormal blocks of vectors P_i and Q_j such that:

$$P_i^H Q_j = I,$$

when $i = j (i, j \leq n)$ and zero otherwise. These vector sets reduce the A matrix to a T_B block tridiagonal matrix form:

$$T_B = \overline{P}_j^H A \overline{Q}_j,$$

where the

$$\overline{P}_j = [P_1, P_2, \ldots P_j]$$

and

$$\overline{Q}_j = [Q_1, Q_2, \ldots Q_j]$$

matrices are the collections of the Lanczos blocks. The structure of the tridiagonal matrix is:

$$T_B = \begin{bmatrix} A_1 & B_2 & & & \\ C_2 & A_2 & . & & \\ & & . & . & . \\ & & & . & . & B_j \\ & & & & C_j & A_j \end{bmatrix}.$$

The bi-orthogonal block Lanczos process is manifested in the following three term recurrence matrix equations:

$$B_{j+1} P_{j+1}^H = P_j^H A - A_j P_j^H - C_j P_{j-1}^H$$

and

$$Q_{j+1} C_{j+1} = AQ_j - Q_j A_j - Q_{j-1} B_j.$$

Note, that in these equations the transpose of the matrix A is avoided.

In order to find the mathematical eigenvalues and eigenvectors we solve two tridiagonal eigenvalue problems posed as:

$$w^H T_J = \mu w^H$$

and

$$T_J z = \mu z,$$

where in the above equations the size of the scalar tridiagonal matrix T_J is $j \times p$, assuming a block size p. It is derived from the block tridiagonal matrix T_B with Givens transformations. The μ eigenvalues of the tridiagonal problem are approximations of the λ eigenvalues of the mathematical problem. The approximations to the eigenvectors of the original problem are calculated from the eigenvectors of the tridiagonal problem by:

$$\underline{y} = \overline{P}_j w$$

and

$$\underline{x} = \overline{Q}_j z,$$

where again $\overline{P}_j, \overline{Q}_j$ are the matrices containing the first j Lanczos blocks of vectors and w, z are the left and right eigenvectors of the tridiagonal problem. Finally, $\underline{x}, \underline{y}$ are the right and left approximated eigenvectors of the mathematical problem.

The useful aspect of the Lanczos method exploited earlier, that the error norm of the original problem may be calculated from the tridiagonal solution without calculating the eigenvectors, is applicable here too. For a similar arrangement, let us introduce a rectangular $n \times p$ matrix E_j having an identity matrix as bottom square. Using this, a residual vector for the left-handed solution is:

$$s^H = \underline{y}^H A - \mu \underline{y}^H = (w^H E_j) B_{j+1} P_{j+1}^H,$$

which means that only the bottom p (if the block size is p) terms of the new eigenvector w are required (due to the structure of E_j). Similarly for the right-handed vectors:

$$r = A\underline{x} - \mu \underline{x} = Q_{j+1} C_{j+1}(E_j^H z).$$

An acceptance criterion (an extension of the one used in the previous chapter) may be based on the norm of above residual vectors as:

$$\min(\frac{||s^H||}{||\underline{y}^H||}, \frac{||r||}{||\underline{x}||}) \leq \epsilon_{\text{acceptance}},$$

where the $\epsilon_{\text{acceptance}}$ value to accept convergence is again given based on the physical problem. The $||.||$ denotes the Euclidean norm.

The physical eigenvalues may easily be recovered from the backward substitution of the spectral transformation.

$$\lambda = \frac{1}{\mu} + \lambda_0.$$

The physical eigenvectors are partitioned as follows.

$$\underline{x} = \begin{bmatrix} \dot{\phi} \\ \phi \end{bmatrix},$$

and

$$\underline{y}^H = \begin{bmatrix} \dot{\psi}^H & \psi^H \end{bmatrix}.$$

10.3 Implicit operator multiplication

The operator multiplication step of the bi-orthogonal Lanczos algorithm is crucial for efficiency. The implicit process described here exploits the structure of the A matrix.

$$A = \begin{bmatrix} -B - M\lambda_0 & -K \\ I & -\lambda_0 I \end{bmatrix}^{-1} \begin{bmatrix} M & 0 \\ 0 & I \end{bmatrix}.$$

It is clear that the matrix is not needed to be built explicitly. In the following the implicit execution of the operator matrix multiplication in both the transpose and non-transpose case is detailed [4].

In the non-transpose case any $z = Ax$ operation in the process will be identical to the

$$\begin{bmatrix} -B - M\lambda_0 & -K \\ I & -\lambda_0 I \end{bmatrix} z = \begin{bmatrix} M & 0 \\ 0 & I \end{bmatrix} x$$

solution of systems of equations. Let us consider the partitioning of the vectors according to the A matrix partitions:

$$\begin{bmatrix} -B - M\lambda_0 & -K \\ I & -\lambda_0 I \end{bmatrix} \begin{bmatrix} z_1 \\ z_2 \end{bmatrix} = \begin{bmatrix} M & 0 \\ 0 & I \end{bmatrix} \begin{bmatrix} x_1 \\ x_2 \end{bmatrix}.$$

Developing the first row of this matrix equation results in

$$(-B - M\lambda_0)z_1 - Kz_2 = Mx_1.$$

Similarly developing the second row, we have

$$z_1 = \lambda_0 z_2 + x_2.$$

Substituting latter into the first row we obtain

$$(-B - M\lambda_0)(\lambda_0 z_2 + x_2) - Kz_2 = Mx_1.$$

Some reordering yields the computational form of

$$-(K + \lambda_0 B + \lambda_0^2 M)z_2 = Mx_1 + (B + \lambda_0 M)x_2.$$

The latter formulation has significant advantages. Besides avoiding the explicit formulation of A, the decomposition of the $2N$ size problem is also avoided.

It is important that the transpose operation be executed without any matrix transpose at all. In this case any $y^T = x^T A$ operation in the process will be identical to the

$$y^T = x^T \begin{bmatrix} -B - M\lambda_0 & -K \\ I & -\lambda_0 I \end{bmatrix}^{-1} \begin{bmatrix} M & 0 \\ 0 & I \end{bmatrix}$$

operation. Let us introduce an intermediate vector z

$$z^T = x^T \begin{bmatrix} -(B + M\lambda_0) & -K \\ I & -\lambda_0 I \end{bmatrix}^{-1}.$$

We partition this vector also according to the matrix partitions and transpose to obtain

$$\begin{bmatrix} x_1 \\ x_2 \end{bmatrix} = \begin{bmatrix} -(B + M\lambda_0)^T & I \\ -K^T & -\lambda_0 I \end{bmatrix} \begin{bmatrix} z_1 \\ z_2 \end{bmatrix}.$$

From the first row of this equation we obtain

$$x_1 = (-B - M\lambda_0)^T z_1 + z_2,$$

and from the second row:

$$x_2 = -K^T z_1 - \lambda_0 z_2.$$

Expressing z_2 from the previous to last equation, substituting into the last and reordering yields

$$x_2 + \lambda_0 x_1 = -(K^T + \lambda_0 B^T + \lambda_0^2 M^T) z_1.$$

From this the computational solution form becomes

$$z_1 = -(K^T + \lambda_0 B^T + \lambda_0^2 M^T)^{-1}(x_2 + \lambda_0 x_1).$$

The lower part of the z vector is recovered from the prior equation as

$$z_2 = x_1 + (B + \lambda_0 M)^T z_1.$$

The latter two equations will also be used in the recovery of the left-handed physical eigenvectors. Finally,

$$y^T = z^T \begin{bmatrix} M & 0 \\ 0 & I \end{bmatrix}.$$

This formulation has even more significant advantages than the non-transpose case. Besides avoiding the explicit formulation of A and the decomposition of the double size problem, the transpose may also be avoided with left handed multiplications and forward-backward substitution.

10.4   Recovery of physical solution

The physical eigenvalues may be recovered from the backward substitution of the complex spectral transformation:

$$\lambda = \frac{1}{\Lambda} + \lambda_0.$$

In order to find the relationship between the mathematical and physical eigenvectors let us write the block form of

$$(\lambda \overline{M} + \overline{K})x = 0,$$

where

$$x = \begin{bmatrix} \dot{\phi} \\ \phi \end{bmatrix}.$$

The block matrices are simply

$$\overline{K} = \begin{bmatrix} B & K \\ -I & 0 \end{bmatrix},$$

and

$$\overline{M} = \begin{bmatrix} M & 0 \\ 0 & I \end{bmatrix}.$$

Substituting and reordering yields

$$[(\overline{K} + \lambda_0 \overline{M})^{-1}\overline{M} + \Lambda I)]x = 0.$$

This proves that the right eigenvectors are invariant under the spectral transformation, i.e. the right physical eigenvectors are the same as their mathematical counterparts, apart from the appropriate partitioning. [5].

For the left handed problem we also use the block notation

$$\underline{y}^H (\lambda \overline{M} + \overline{K}) = 0$$

with a left handed physical eigenvector of

$$\underline{y}^H = \begin{bmatrix} \dot{\psi}^H & \psi^H \end{bmatrix}.$$

Substituting again and introducing an appropriate identity matrix to accommodate the left multiplication yields

$$\underline{y}^H (\overline{K} + \lambda_0 \overline{M})(\overline{K} + \lambda_0 \overline{M})^{-1}[\overline{M} + (\overline{K} + \lambda_0 \overline{M})\Lambda] = 0.$$

By multiplying we obtain

$$\underline{y}^H (\overline{K} + \lambda_0 \overline{M})[(\overline{K} + \lambda_0 \overline{M})^{-1}\overline{M} + \Lambda I] = 0,$$

which is equivalent to

$$[(\overline{K} + \lambda_0 \overline{M})^H \underline{y}]^H [(\overline{K} + \lambda_0 \overline{M})^{-1}\overline{M} + \Lambda I] = 0.$$

Since the original mathematical problem we solve is

$$y^H [(\overline{K} + \lambda_0 \overline{M})^{-1}\overline{M} + \Lambda I] = 0,$$

the left handed physical eigenvectors are not invariant under the transformation. Comparing gives the following relationship:

$$y^H = [(\overline{K} + \lambda_0 \overline{M})^H \underline{y}]^H,$$

or expanding into the solution terms

$$y = -\begin{bmatrix} -B - M\lambda_0 & -K \\ I & -\lambda_0 I \end{bmatrix}^H \underline{y}.$$

Finally

$$\underline{y} = -\begin{bmatrix} -B - M\lambda_0 & -K \\ I & -\lambda_0 I \end{bmatrix}^{-1,H} y.$$

The cost of this backtransformation is not very large since the factors of the operator matrix are available; we need a forward-backward substitution only.

10.5   Solution evaluation

Since the mathematical solution is significantly different from the physical solution, some additional accuracy considerations are needed. From the eigenvalue solution it will be guaranteed that the left and right mathematical eigenvectors are orthonormal to computational accuracy.

$$Y^H X = \overline{I}.$$

Here $\overline{I}$ is an approximate identity matrix with off-diagonals as computational zeroes. Monitoring these terms and printing the largest one or the ones above a certain threshold will give a final orthogonality check.

Based on the physical eigenvalues recovered by the shift formulae and the physical eigenvectors another orthogonality criterion can be formed. Using the left and right solutions, the following equations are true for the problem:

$$(M\lambda_i^2 + B\lambda_i + K)\phi_i = 0,$$

and

$$\psi_j^H(M\lambda_j^2 + B\lambda_j + K) = 0.$$

By appropriate pre- and post-multiplications we get

$$\psi_j^H(M\lambda_i^2 + B\lambda_i + K)\phi_i = 0,$$

and

$$\psi_j^H(M\lambda_j^2 + B\lambda_j + K)\phi_i = 0.$$

A subtraction yields

$$\psi_j^H M \phi_i (\lambda_i^2 - \lambda_j^2) + \psi_j^H B \phi_i (\lambda_i - \lambda_j) = 0.$$

Assuming $\lambda_j \neq \lambda_i$ we can shorten and obtain an orthogonality criterion as:

$$O_1 = \psi_j^H M \phi_i (\lambda_i + \lambda_j) + \psi_j^H B \phi_i.$$

The O_1 matrix also has computational zeroes as off diagonal terms (when $i \neq j$) and nonzero (containing $2\lambda_i$) diagonal terms.

Now premultiplying by $\lambda_j \psi_j^H$, postmultiplying by $\lambda_i \phi_i$ and subtracting we obtain

$$\lambda_j \psi_j^H (M\lambda_i^2 + B\lambda_i + K)\phi_i - \psi_j^H (M\lambda_j^2 + B\lambda_j + K)\lambda_i \phi_i = 0.$$

By expanding and canceling we get

$$(\lambda_i - \lambda_j)\psi_j^H M \phi_i \lambda_i \lambda_j + (\lambda_j - \lambda_i)\psi_j K \phi_i = 0.$$

Assuming again that $\lambda_j \neq \lambda_i$ we can shorten and obtain another orthogonality criterion recommended mainly for the structural damping option (the case when there is no B matrix, but damping is present in the K matrix) as

$$O_2 = \lambda_j \psi_j^H M \lambda_i \phi_i - \psi_j^H K \phi_i.$$

This orthogonality matrix will also have zero off-diagonal terms, but nonzero (containing λ_i^2) diagonal terms.

10.6 Reduction to Hessenberg form

For dense, unsymmetric matrices the Householder reduction technique of the last chapter is still applicable, but will result in an upper Hessenberg form. Let us review this transformation in detail. Due to the lack of symmetry, the pre- and post-multiplications must be considered separately. First we pre-multiply

$$P_r A_{r-1} = (I - \beta v_r v_r^T) A_{r-1} = A_{r-1} - \beta v_r (v_r A_{r-1}) = B_r.$$

Because v_r has zeroes in its first $r - 1$ rows, the pre-multiplication leaves the first $r - 1$ rows of A_{r-1} unchanged. The post-multiplication of

$$A_r = B_r P_r = B_r (I - \beta v_r v_r^T) = B_r - \beta (B_r v_r) v_r^T$$

will leave the first $r - 1$ columns of A_{r-1} unchanged.

This may be computationally efficiently executed by the following steps:

1. $p_r^T = v_r^T A_{r-1}$

2. $B_r = A_{r-1} - 2v_r p_r^T$

3. $q_r = B_r v_r$

4. $A_r = B_r - 2q_r v_r^T$

The latter steps enable the execution of this computation in the case when the complete A matrix does not reside in memory, a common occurrence in the industry.

Finally, after $n - 2$ transformations we obtain

$$A_{n-2} = H,$$

where H is a matrix in upper Hessenberg form, having all terms below the sub-diagonal zero:

$$H = \begin{bmatrix} x & x & x & x & x \\ x & x & x & x & x \\ 0 & x & x & x & x \\ 0 & 0 & x & x & x \\ 0 & 0 & 0 & x & x \end{bmatrix}.$$

This reduction is again a congruence transformation. The eigenvalue calculation method for tridiagonal matrices introduced in Section 9.4 is also applicable here, however, the final result is an upper triangular matrix, as opposed to a diagonal matrix.

10.7   Rotating component application

An important engineering application resulting in the need for complex modal reduction is the analysis of rotating components. The gyroscopic effects are modeled by the Coriolis matrix that is related to the velocity term of the analysis. The Coriolis matrix of a single node point for a rotation about the z axis is formed as

$$C_i = \begin{bmatrix} 0 & -m & 0 & 0 & 0 & 0 \\ m & 0 & 0 & 0 & 0 & 0 \\ 0 & 0 & 0 & 0 & 0 & 0 \\ 0 & 0 & 0 & 0 & I & 0 \\ 0 & 0 & 0 & -I & 0 & 0 \\ 0 & 0 & 0 & 0 & 0 & 0 \end{bmatrix},$$

where

$$I = \frac{1}{2}(I_{zz} - (I_{xx} + I_{yy})),$$

and the I_{**} terms are the second order moments of inertia to the respective axis. The C_i matrices are assembled into the global Coriolis matrix C [6]. For a symmetric rotor, a most practical application, the inertia term simplifies, since

$$I_{xx} = I_{yy}.$$

Furthermore, the centripetal force results in a phenomenon called the centrifugal softening and the matrix describing this for a single node for rotation about the z axis results in

$$H_i = \begin{bmatrix} m & 0 & 0 & 0 & 0 & 0 \\ 0 & m & 0 & 0 & 0 & 0 \\ 0 & 0 & 0 & 0 & 0 & 0 \\ 0 & 0 & 0 & -I_{zz} + I_{yy} & 0 & 0 \\ 0 & 0 & 0 & 0 & -I_{zz} + I_{xx} & 0 \\ 0 & 0 & 0 & I & 0 & 0 \\ 0 & 0 & 0 & 0 & 0 & 0 \end{bmatrix},$$

where the m is the mass associated with the single node. These matrices are also assembled into the global centrifugal softening matrix H.

There are three main sources of damping in a rotating system. The internal damping of the rotor and the external damping acting on the bearing populate the damping matrix B. The third damping source is the structural damping related to the displacement, and as such appearing as an anti-symmetric stiffness matrix component K_b.

With these the free vibration of a rotating structure in a rotating reference system is described by [3]

$$(\lambda^2 M + i\lambda(B + 2\Omega C) + (K + K_b - \Omega^2 H))\phi = 0.$$

The matrices M, B and K are the regular mass, damping and stiffness matrices of earlier chapters. The C Coriolis matrix is anti-symmetric, but the H

centrifugal softening matrix is symmetric.

All matrices are real, except for the stiffness matrix K which may be complex if there is structural damping also applied to the model.

The Ω is the rotational speed and the eigenvalues appear in complex conjugate pairs as

$$\lambda = \alpha \pm i\omega.$$

The natural frequency of the mode shapes is

$$f = \frac{\omega}{2\Pi},$$

and the damping coefficient is

$$g = \frac{2\alpha}{\omega}.$$

Plotting the frequencies as function of the rotation speed results in the Campbell diagram [2], such as shown in Figure 10.1. The figure shows the results translated into a fixed reference system since it is more interpretive for the engineer. The translation between the fixed and rotational frame of reference is by simply adding or subtracting the rotor speed from the curves.

The $1P$ (one per revolution) line starting from the origin in the figure represents the locations of equal frequency and rotor speed. Intersections of the modes with this curve pin-point the critical speed locations. Those locations should be avoided in the operational range of the rotor to avoid resonance catastrophes.

The example in the figure has four modes plotted. The first three modes intersect the $1P$ line resulting in three critical speeds. The negative slope curves of modes $1, 3$ represent backward whirl and the positive slope curve of modes $2, 4$ represents a forward whirl motion. The forward whirl motion, also called direct whirl, is in the direction of the rotation, while the backward or reverse whirl is opposite.

The whirl direction may be computed from the eigenvectors as follows. The real and imaginary components of every node point gathered into two vectors whose ith terms are

$$\Psi_{Re}(i) = \begin{bmatrix} \phi_x^{Re}(i) \\ \phi_y^{Re}(i) \\ 0 \end{bmatrix},$$

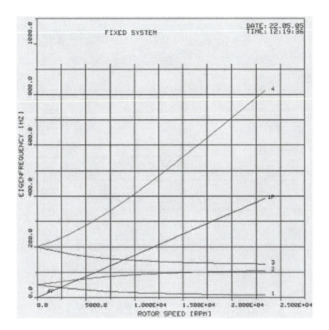

FIGURE 10.1 Campbell diagram

and

$$\Psi_{Im}(i) = \begin{bmatrix} \phi_x^{Im}(i) \\ \phi_y^{Im}(i) \\ 0 \end{bmatrix}.$$

The whirl direction is forward if the vector

$$w = \Psi_{Re} \times \Psi_{Im}$$

points to the positive z-direction, backwards otherwise.

By plotting the real parts of the eigenvalues as function of the rotation speed, such as shown in Figure 10.2, the regions of instability of the rotor may be established. Unstable regions are where the real part of any mode is positive. In the case of this example, the region of instability starts at approximately 5, 000 RPM where mode 2 crosses the horizontal axis. Both of these graphs and the related computations are very important for the engineer analyzing rotational machinery.

The fact that these graphs were plotted in the fixed frame of reference while computed in a rotating frame is noteworthy. The analysis of rotating

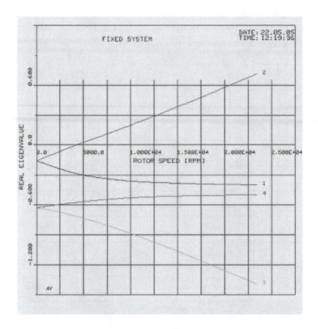

FIGURE 10.2 Stability diagram

components may also be directly executed in a fixed reference system. In this case, however, Steiner inertia terms of the form:

$$\frac{1}{2}mr^2,$$

need to be added to the matrices, and the rotor must be symmetric. For unsymmetric rotor models, such as helicopter rotors, propellers and wind-mills, using the rotational reference system is necessary.

10.8   Complex modal analysis case studies

As first application example we consider another automobile component, the brake. The physical phenomenon of braking is a stick-slip type vibration between the friction pads on the caliper and the rotor.

The friction between the pads and the rotor is a function of both pressure and velocity. The pressure-based friction, sometimes called friction stiffness, creates a relationship between the normal and tangential components of the brake elements. The mathematical effect of this is the asymmetry of the K stiffness matrix of the finite element model. The velocity-based friction, also known as friction damping introduces the B damping matrix of the model.

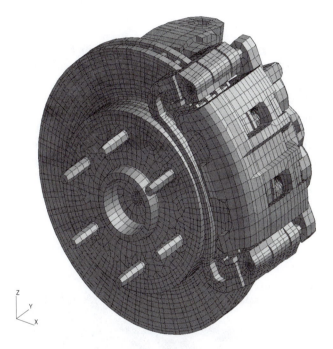

FIGURE 10.3 Brake model

A brake model of a test problem (such as shown in Figure 10.3.) had the model and execution statistics shown on Table 10.1.

The complex eigenvalue analysis of this brake model was executed with the block, bi-orthogonal Lanczos method described above. The workstation used had 1.5 GHz clock speed and the elapsed time shown in the table is in minutes:seconds.

The statistics demonstrate the time-consuming nature of the complex modal calculation. In essence, the expense of computing complex modes of an unsymmetric problem is four-fold. In the normal modes analysis case study in

TABLE 10.1
Statistics of brake model

Model	Nodes	Elements	g-size	f-size
	50,305	33,750	301,833	298,724
Execution	Modes	CPU-sec	I/O-GB	Elapsed
	190	5,098.6	393.4	86:35

Section 9.6 we have computed a little more than four times as many eigen-vectors of a seven times larger problem in about the same time, using eight processors.

Higher node and element size models commonly occur in the rotating ma-chinery industry. A part of such a model is shown in Figure 10.4 and the partial view into the cylinder from the left side hints at a difficult interior.

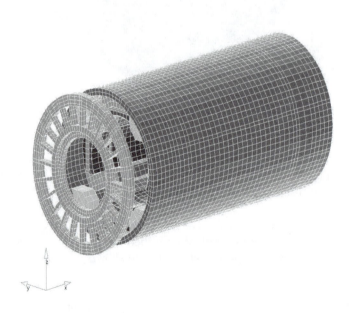

FIGURE 10.4 Rotating machinery model

Table 10.2 shows the matrix statistics of the rotating machinery model. Note that the stiffness matrix was complex. The "max terms" represents maximum number of terms in the densest column of the matrix. The automatically generated, very detailed model consisted of 213,000 nodes and 130,000 elements. This model clearly demonstrates the computational complexity of rotating dynamic applications.

TABLE 10.2
Complex eigenvalue analysis statistics

K	number of rows	nonzero terms	max terms
matrix	638,057	24.8 million	405
M	number of rows	zero columns	max terms
matrix	638,057	368	1
Factor	number of rows	nonzero terms	max front
matrix	638,057	670 million	7,668

The implicit solution technique detailed in this chapter and the block unsymmetric bi-orthogonal method were used in the complex eigenvalue analysis of this model. The analysis took 250 elapsed minutes on a workstation containing 4 (1.5 GHz) CPUs. Of this about 170 minutes were spent in the complex eigenvalue analysis computation extracting the complex eigenpairs at each rotational frequency.

The amount of I/O was 1.2 GBytes and the disk footprint was 16 Gigabytes. This is mainly due to the fact that the structural matrices were residing out of core, a commonplace phenomenon in industry.

References

[1] Bai, Z. et al; ABLE: an adaptive block Lanczos method for non-Hermitian eigenvalue problems, SIAM Journal on Matrix Analysis and Applications, Vol. 20, pp. 1060-1082, 1999

[2] Campbell, W.; Protection of steam turbine disk wheels from axial vibration, Transactions of ASME, No. 46, pp.. 31-160, 1924

[3] Genta, G.; Dynamics of rotating systems, Springer, New York, 2005

[4] Komzsik, L.; Implicit solution of quadratic eigenvalue problems, Finite Elements in Analysis and Design, Vol 35., pp. 799-810, Elsevier, 2001

[5] Komzsik, L.; The Lanczos method: Evolution and Application, SIAM, Philadelphia, 2003

[6] Vollan, A.; GAROS reference manual, AeroFEM GmbH, 1995

11

Dynamic Reduction

In this chapter, we formulate a reduction technique applicable to dynamic analysis problems. This is a demanding technique and only approximate solutions are produced. However, by the virtue of the reduction, the actual computational complexity may be reduced. The relevant publications listed in the reference range from the classical [2] to the recent [4] and from the theoretical to the practical [9].

We discuss the procedure in the context of undamped, free vibration or normal modes problem discussed in the previous to last chapter. This analysis problem is described as

$$[K_{aa} - \lambda M_{aa}]u_a = 0,$$

where M_{aa} and K_{aa} are the a partition mass and stiffness matrices, and u_a is the eigenvector of the structure corresponding to eigenvalue λ.

11.1   Single-level, single-component dynamic reduction

Following the single-level, single-component static condensation principle, the normal modes problem may also be partitioned into interior and boundary components.

$$\begin{bmatrix} K_{oo} - \lambda M_{oo} & K_{ot} - \lambda M_{ot} \\ K_{ot}^T - \lambda M_{ot}^T & K_{tt} - \lambda M_{tt} \end{bmatrix} \begin{bmatrix} u_o \\ u_t \end{bmatrix} = 0.$$

Let us first solve the following interior eigenvalue problem of

$$[K_{oo} - \lambda M_{oo}]\phi_o = 0.$$

We construct a diagonal matrix $\Lambda_{q_o q_o}$ containing q_o eigenvalues and a rectangular $\Phi_{o q_o}$ matrix that contains all corresponding ϕ_o eigenvectors. The latter are called the fixed boundary interior modes, as this equation assumes that the boundary degrees of freedom have zero displacement, i.e., the boundary is fixed. They are usually, but not necessarily, chosen to be mass-orthogonal, so the so-called generalized mass is identity:

$$\Phi_{oq_o}^T M_{oo} \Phi_{oq_o} = I_{q_o q_o}.$$

The generalized stiffness yields the eigenvalues

$$\Phi_{oq_o}^T K_{oo} \Phi_{oq_o} = \Lambda_{q_o q_o}.$$

The number of eigenvalues extracted from the interior problem is q_o and it is usually much less than the o partition size. This is a result of a compromise. On one hand, the numerical accuracy of the solution increases with higher number of interior modes. On the other hand, reducing the problem size by retaining fewer than all the modes, produces the computational advantage.

In the industry, the number of eigenvalues extracted from the o partition depends on the width of the frequency range in which the global modes are sought. A heuristic value of 1.5 times the width of the global frequency range is used to set the local frequency range of interest. The accuracy of the reduction is addressed in more detail in the next section.

Similarly, the eigenvalue problem of the boundary

$$[K_{tt} - \lambda M_{tt}]\phi_t = 0$$

produces the so-called boundary mode shapes, gathered into Φ_{tq_t}. In this case, q_t refers to the number of eigenvalues found in the boundary problem and $\Lambda_{q_t q_t}$ is a diagonal matrix containing those eigenvalues. The dynamic reduction transformation matrix is formed as

$$S = \begin{bmatrix} \Phi_{oq_o} & \\ & \Phi_{tq_t} \end{bmatrix}.$$

Note, that this matrix is rectangular as the number of mode shapes extracted from the interior and from the boundary are both less than the size of the partition. S has $d = q_t + q_o$ columns and $t + o$ rows. Pre-multiplying by S^T and substituting

$$Su_d = u_a$$

yields

$$S^T[K_{aa} - \lambda M_{aa}]Su_d = 0.$$

Executing the multiplications produces a dynamically reduced problem of

$$[K_{dd} - \lambda M_{dd}]u_d = 0.$$

The reduced problem in detail is

$$\begin{bmatrix} \Lambda_{q_o q_o} - \lambda I_{q_o q_o} & K_{q_o q_t} - \lambda M_{q_o q_t} \\ K_{q_o q_t}^T - \lambda M_{q_o q_t}^T & \Lambda_{q_t q_t} - \lambda I_{q_t q_t} \end{bmatrix} \begin{bmatrix} u_{q_o} \\ u_{q_t} \end{bmatrix} = 0.$$

Here the transformed coupling matrices are

$$M_{q_o q_t} = \Phi_{o q_o}^T M_{ot} \Phi_{t q_t},$$

and

$$K_{q_o q_t} = \Phi_{o q_o}^T K_{ot} \Phi_{t q_t}.$$

The transformed eigenvector components are defined by

$$\Phi_{o q_o} u_{q_o} = u_o,$$

and

$$\Phi_{t q_t} u_{q_t} = u_t.$$

The recovery of the solution of the original size eigenvector (which is the subject of the engineer's interest) is simply the application of the last two equations.

The reduced problem's size is $q_o + q_t$ and has a very specific structure. The matrices are diagonal apart from the coupling terms. Clearly there are computational advantages to be gained by solving this reduced problem. On the other hand, there is a truncation error introduced which will be addressed in detail in next section as well as in Chapter 13.

The dynamic behavior of the finite element model may be (computationally) exactly reproduced if the dynamic reduction executed on the stiffness and mass matrices is done with all the eigenvectors, the case of the full reduction. There may still be a computational advantage in calculating the response of a system with the matrices of the full reduction, as they are much sparser, even though they are of the same dimensions as the original matrices.

11.2 Accuracy of dynamic reduction

While our focus has been and remains on the computational side of these techniques, a few words on the accuracy of dynamic reduction are warranted. Recall the original analysis problem of

$$K_{aa} u_a = \lambda_a M_{aa} u_a.$$

The transformations of

$$K_{dd} = S^T K_{aa} S,$$
$$M_{dd} = S^T M_{aa} S,$$

and

$$u_a = Su_d$$

resulted in the reduced problem:

$$K_{dd}u_d = \lambda_d M_{dd} u_d.$$

Note the distinction, made between the eigenvalues computed from the two problems for this discussion. The eigensolution of the reduced problem is an approximation from the subspace $\mathcal{S}$ spanned by S. Let us evaluate the accuracy of eigensolution obtained from the reduced problem.

Assume that the analytical eigenvalues of the unreduced problem are

$$\lambda_{a1} \leq \lambda_{a2} \leq \ldots \leq \lambda_{an}$$

and the eigenvalues of the reduced problem are

$$\lambda_{d1} \leq \lambda_{d2} \leq \ldots \leq \lambda_{dk}.$$

Here n is the number of eigenvalues of the unreduced problem and k is that of the reduced problem. The value of k is the dimension of the subspace from which the approximate solution is obtained, specifically

$$k = q_o + q_t.$$

Let us define the norm of a vector x with respect to the mass matrix M as

$$||x||_M = \sqrt{x^T M x}$$

and the angle between the vector and a subspace $\mathcal{S}$ spanned by S in an M-inner product space as

$$cos\angle(x, \mathcal{S})_M = \frac{||S^T M x||_2}{||x||_M}.$$

Here we used the assumption of

$$S^T M S = I.$$

Based on these it is possible to give bounds for the solutions of the reduced problem [7]. For the eigenvalues the

$$\lambda_{d1} - \lambda_{a1} \leq (\lambda_{an} - \lambda_{a1})sin^2\angle(u_{a1}, \mathcal{S})_M$$

inequality holds. In this context the M matrix is the M_{aa} matrix. This means that the error between the eigenvalue of the reduced problem and the corresponding ("exact") eigenvalue of the unreduced problem is bound by the angle between the corresponding eigenvector and the subspace from which the reduced solution was obtained.

In case of a full reduction, $q_o = o, q_t = t$ the subspace has the same dimension as the original space, $k = n$ and the angle between the eigenvector and the subspace is zero. This implies that the reduced problem solution is computationally exact. The reduction is then a congruence transformation: since S is nonsingular, the eigenvalues of the (K_{aa}, M_{aa}) matrix pencil are identical to those of the (K_{dd}, M_{dd}) pencil.

The $\lambda_{an} - \lambda_{a1}$ term on the right-hand side represents the eigenspectrum of the problem. This implies that the accuracy of the dynamic reduction also depends on the segment of the eigenspectrum represented by the component eigenvectors. Usually the lower end of the spectrum up to a certain frequency is of practical interest, however, with the help of the spectral transformation described in Section 9.1, the focus may be placed on the upper end or the middle of the spectrum.

For the error between the eigenvector from the reduced problem and the "exact" eigenvector of the unreduced problem, the bound is

$$sin^2 \angle (\overline{u}_{d1}, u_{a1})_M \leq \sqrt{\frac{\lambda_{an} - \lambda_{a1}}{\lambda_{a2} - \lambda_{a1}}} sin \angle (u_{a1}, \mathcal{S})_M.$$

Here $\overline{u}_{d1}$ is an appropriately inflated a-size version of u_{d1}. The $\lambda_{a2} - \lambda_{a1}$ term is indicative of the spacing of the eigenvalues. In both bounds the angle term may be computed by the definition as

$$sin \angle (u_{a1}, \mathcal{S})_M = \sqrt{1 - (\frac{||S^T M x||_2}{||x||_M})^2}.$$

For a large problem and large subspace size the computation between the eigenvector and the subspace may be simplified to

$$sin \angle (u_{a1}, \mathcal{S})_M = sin \angle (u_{a1}, v_1)_M,$$

where the v_1 vector is the vector closest to the eigenvector in the M-inner product space

$$v_1 : min_{v \in \mathcal{S}} ||u_{a1} - v||_M.$$

These bounds are *a posteriori* in nature. There is ongoing research and there are some recent results in providing *a priori* bounds., i.e., in terms of the reduced problem's solution, see [8] for example. These methods are not widely accepted yet in the industry and will not be discussed further here. The accuracy affect of the dynamic reduction to the response of the system will be addressed in more detail in Chapter 13.

11.3   Computational example

For a dynamic reduction example let us use as the stiffness matrix the result
of the static condensation example of Section 8.2.

$$\overline{K} = \begin{bmatrix} 2 & 1 & 0 \\ 1 & 3 & 0 \\ 0 & 0 & 18/5 \end{bmatrix},$$

and as mass matrix

$$\overline{M} = \begin{bmatrix} 1 & 1/2 & 0 \\ 1/2 & 1 & 1/5 \\ 0 & 1/5 & 18/25 \end{bmatrix}.$$

The analytic solution of the

$$(\overline{K} - \lambda \overline{M})\overline{u} = 0$$

problem is

$$\Lambda = \begin{bmatrix} 2 & 0 & 0 \\ 0 & 3 & 0 \\ 0 & 0 & 6 \end{bmatrix}$$

and

$$\overline{U} = \begin{bmatrix} 1 & -6 & 3 \\ 0 & 12 & -6 \\ 0 & 5 & 10 \end{bmatrix}.$$

Here Λ is a diagonal matrix containing all three eigenvalues and $\overline{U}$ contains
the eigenvectors.

We assume the same partitioning as used in Section 8.2, namely that the
upper left, in this case 2 by 2 size, partition is the interior. The eigenvalues
of the interior partition are

$$\Lambda_{q_o q_o} = \begin{bmatrix} 2 & 0 \\ 0 & 5/2 \end{bmatrix},$$

and the eigenvectors

$$\Phi_{o q_o} = \begin{bmatrix} 1 & -1/2 \\ 0 & 1 \end{bmatrix}.$$

The generalized mass matrix with these (not mass normalized) eigenvectors is

$$m_{q_o q_o} = \begin{bmatrix} 1 & 0 \\ 0 & 3/4 \end{bmatrix}.$$

The mass coupling term is

$$M_{qt} = \begin{bmatrix} 1 & 0 \\ -1/2 & 1 \end{bmatrix}^T \begin{bmatrix} 0 \\ 1/5 \end{bmatrix} = \begin{bmatrix} 0 \\ 1/5 \end{bmatrix}.$$

The eigensolution components of the boundary modal solution are

$$\Lambda_{q_t q_t} = \begin{bmatrix} 18/5 \end{bmatrix},$$

and

$$\Phi_{t q_t} = \begin{bmatrix} 1 \end{bmatrix}.$$

The boundary generalized mass is

$$m_{q_t q_t} = \begin{bmatrix} 18/25 \end{bmatrix}.$$

Hence, the transformation matrix of the dynamic reduction is

$$S = \begin{bmatrix} 1 & -1/2 & 0 \\ 0 & 1 & 0 \\ 0 & 0 & 1 \end{bmatrix}.$$

Note, that since the number of eigenvectors extracted from the interior is the same as the size of the interior ($q_o = o = 2$), this matrix is now square. With this, the matrices of the dynamically reduced eigenproblem are

$$K_{dd} = \begin{bmatrix} 2 & 0 & 0 \\ 0 & 5/2 & 0 \\ 0 & 0 & 18/5 \end{bmatrix}$$

and

$$M_{dd} = \begin{bmatrix} 1 & 0 & 0 \\ 0 & 3/4 & 1/5 \\ 0 & 1/5 & 18/25 \end{bmatrix}.$$

The reduced

$$(K_{dd} - \lambda M_{dd})u_d = 0$$

problem's eigenvalues are

$$\Lambda = \begin{bmatrix} 2 & 0 & 0 \\ 0 & 3 & 0 \\ 0 & 0 & 6 \end{bmatrix}.$$

These eigenvalues are identical to the eigenvalues of the $(\overline{K} - \lambda \overline{M})\overline{u} = 0$ analytic global solution as the transformation is a congruence transformation. The eigenvectors are

$$U_{dd} = \begin{bmatrix} 1 & 0 & 0 \\ 0 & 1 & -3/5 \\ 0 & 5/12 & 1 \end{bmatrix} = \begin{bmatrix} U_{od} \\ U_{td} \end{bmatrix}.$$

To obtain the global eigenvector solution, we need to recover the effects of dynamic reduction. For the interior:

$$\overline{U}_o = \Phi_{oq_o} U_{od} = \begin{bmatrix} 1 & -1/2 \\ 0 & 1 \end{bmatrix} \begin{bmatrix} 1 & 0 & 0 \\ 0 & 1 & -3/5 \end{bmatrix} = \begin{bmatrix} 1 & -1/2 & 3/10 \\ 0 & 1 & -3/5 \end{bmatrix}.$$

For the boundary

$$\overline{U}_t = \Phi_{tq_t} U_{td} = \begin{bmatrix} 1 \end{bmatrix} \begin{bmatrix} 0 & 5/12 & 1 \end{bmatrix} = \begin{bmatrix} 0 & 5/12 & 1 \end{bmatrix}.$$

Hence, the recovered global eigenvectors of our problem are

$$\overline{U} = \begin{bmatrix} 1 & -1/2 & 3/10 \\ 0 & 1 & -3/5 \\ 0 & 5/12 & 1 \end{bmatrix}.$$

These eigenvectors are the same as the analytic global solution apart from a scalar multiplier.

11.4   Single-level, multiple-component dynamic reduction

As in the case of static condensation, dynamic reduction may also be expanded to multiple components. In this case, the partitioned eigenvalue problem is

$$\begin{bmatrix} K^1_{oo} - \lambda M^1_{oo} & & & K^1_{ot} - \lambda M^1_{ot} \\ & \ddots & & \vdots \\ & & K^i_{oo} - \lambda M^i_{oo} & K^i_{ot} - \lambda M^i_{ot} \\ & & & \vdots \\ K^{T,1}_{ot} - \lambda M^{T,1}_{ot} & \cdots & K^{T,i}_{ot} - \lambda M^{T,i}_{ot} & \cdots & K_{tt} - \lambda M_{tt} \end{bmatrix} \begin{bmatrix} u^1_o \\ \vdots \\ u^i_o \\ \vdots \\ u_t \end{bmatrix} = 0.$$

For each component, fixed boundary interior mode shapes are computed as

$$K^i_{oo} \Phi^i_{oq^i_o} = M^i_{oo} \Phi^i_{oq^i_o} \Lambda^i_{q^i_o q^i_o}.$$

The coupled boundary mode shapes are computed the same way as before

$$K_{tt} \Phi_{tq_t} = M_{tt} \Phi_{tq_t} \Lambda_{q_t q_t}.$$

The dynamic reduction transformation matrix becomes

$$
S = \begin{bmatrix}
\Phi^1_{oq^1_o} & & & \\
& \ddots & & \\
& & \Phi^i_{oq^i_o} & \\
& & & \ddots \\
& & & & \Phi_{tq_t}
\end{bmatrix},
$$

where the q^i_o index denotes the number of eigenvectors from the interior (o partition) of the i-th component. Note, that this matrix is also rectangular as the number of mode shapes extracted from each component is very likely less than the size of the component. S has $d = q_t + \Sigma^n_{i=1} q^i_o$ columns and $t + \Sigma^n_{i=1} o^i$ rows. Here n is the number of components. Pre-multiplying by S^T and substituting

$$
S u_d = u_a
$$

yields

$$
S^T [K_{aa} - \lambda M_{aa}] S u_d = 0.
$$

Executing the multiplications results in

$$
[K_{dd} - \lambda M_{dd}] u_d = 0.
$$

The detailed structure of the reduced problem is

$$
\begin{bmatrix}
\Lambda^1_{q^1_o q^1_o} - \lambda I^1_{q^1_o q^1_o} & & & & K^1_{q^1_o q_t} - \lambda M^1_{q^1_o q_t} \\
& \ddots & & & \vdots \\
& & \Lambda^i_{q^i_o q^i_o} - \lambda I^i_{q^i_o q^i_o} & & K^1_{q^i_o q_t} - \lambda M^i_{q^i_o q_t} \\
& & & \ddots & \vdots \\
K^{1,T}_{q^1_o q_t} - \lambda M^{1,T}_{q^1_o q_t} & \cdots & K^{i,T}_{q^i_o q_t} - \lambda M^{i,T}_{q^i_o q_t} & \cdots & \Lambda_{q_t q_t} - \lambda I_{q_t q_t}
\end{bmatrix}
\begin{bmatrix}
u^1_{q^1_o} \\
\vdots \\
u^i_{q^i_o} \\
\vdots \\
u_{q_t}
\end{bmatrix} = 0.
$$

The reduced problem's size is now $d = q_t + \Sigma^n_{i=1} q^i_o$ and still has a very specific structure. Its diagonal blocks are the generalized stiffness and mass matrices that are diagonal, so the complete reduced matrix is diagonal apart from the coupling block of the sides.

The diagonal generalized stiffnesses are

$$
\Lambda^i_{q^i_o q^i_o} = \Phi^{i,T}_{oq^i_o} K^i_{oo} \Phi^i_{oq^i_o},
$$

and

$$
\Lambda_{q_t q_t} = \Phi^T_{tq_t} K_{tt} \Phi_{tq_t}.
$$

Assuming mass-orthogonality of the eigenvectors, the generalized mass matrices are identity

$$
I^i_{q^i_o q^i_o} = \Phi^{i,T}_{oq^i_o} M^i_{oo} \Phi^i_{oq^i_o},
$$

and

$$I_{q_t q_t} = \Phi_{t q_t}^T M_{tt} \Phi_{t q_t}.$$

The mass and stiffness coupling terms are

$$M_{q_o^i q_t}^i = \Phi_{o q_o^i}^{i,T} M_{ot}^i \Phi_{t q_t}$$

and

$$K_{q_o^i q_t} = \Phi_{o q_o^i}^{i,T} K_{ot}^i \Phi_{t q_t}.$$

The transformed eigenvector components are defined by

$$\Phi_{o q_o^i}^i u_{q_o^i}^i = u_o^i$$

and

$$\Phi_{t q_t} u_{q_t} = u_t.$$

These equations are the basis for the engineering solution in the case of single level, multiple-component dynamic reduction.

11.5   Multiple-level dynamic reduction

It is possible to formulate the dynamic reduction scheme in multiple levels also. We are still addressing the problem of

$$[K_{aa} - \lambda M_{aa}] u_a = 0.$$

Let us assume the same 2 levels, 2 components per level partitioning of Section 8.4. Then the mass matrix is partitioned as

$$M_{aa} = \begin{bmatrix} M_{oo}^1 & & M_{ot}^{1,12} & & & & M_{ot}^{1,0} \\ & M_{oo}^2 & M_{ot}^{2,12} & & & & M_{ot}^{2,0} \\ & & M_{tt}^{12} & & & & M_{tt}^{12,0} \\ & & & M_{oo}^3 & & M_{ot}^{3,34} & M_{ot}^{3,0} \\ & sym & & & M_{oo}^4 & M_{ot}^{3,34} & M_{ot}^{4,0} \\ & & & & & M_{tt}^{34} & M_{tt}^{34,0} \\ & & & & & & M_{tt}^0 \end{bmatrix}.$$

The dynamic reduction transformation matrix is a collection of the interior component mode shapes and the boundary mode shapes.

$$S = \begin{bmatrix} \Phi^1_{oq^1_o} & & & & & & \\ & \Phi^2_{oq^2_o} & & & & & \\ & & \Phi^{12}_{tq_{t12}} & & & & \\ & & & \Phi^3_{oq^3_o} & & & \\ & & & & \Phi^4_{oq^4_o} & & \\ & & & & & \Phi^{34}_{tq_{t34}} & \\ & & & & & & \Phi^0_{tq^0_t} \end{bmatrix}$$

where now the q_{t12} index denotes the number of eigenvectors from the boundary of the first and second partitions. Similarly the q_{t34} index denotes the number of eigenvectors from the boundary of the third and fourth partition and q_{t0} index denotes the number of eigenvectors of the final boundary.

Note, that this matrix is also rectangular as the number of mode shapes extracted from each component is very likely less than the size of the component. S has $d = q_{t12} + q_{t34} + q_{t0} + \Sigma^n_{i=1} q^i_o$ columns and the original number of rows. Pre-multiplying by S^T and substituting

$$Su_d = u_a$$

yields

$$S^T[K_{aa} - \lambda M_{aa}]Su_d = 0.$$

Executing the multiplications result in

$$[K_{dd} - \lambda M_{dd}]u_d = 0.$$

The detailed structure of the reduced stiffness matrix is

$$K_{dd} = \begin{bmatrix} \Lambda^1_{q^1_o q^1_o} & & K_{q^1_o q_{t12}} & & & & K_{q^1_o q_{t0}} \\ & \Lambda^2_{q^2_o q^2_o} & K_{q^2_o q_{t12}} & & & & K_{q^2_o q_{t0}} \\ & & \Lambda^{12}_{q_{t12} q_{t12}} & & & & K_{q_{t12} q_{t0}} \\ & & & \Lambda^3_{q^3_o q^3_o} & & K_{q^3_o q_{t34}} & K_{q^3_o q_{t0}} \\ & sym & & & \Lambda^4_{q^4_o q^4_o} & K_{q^4_o q_{t34}} & K_{q^4_o q_{t0}} \\ & & & & & \Lambda^{34}_{q_{t34} q_{t34}} & K_{q_{t34} q_{t0}} \\ & & & & & & \Lambda^0_{q_{t0} q_{t0}} \end{bmatrix}.$$

The detailed structure of the reduced mass matrix is

$$M_{dd} = \begin{bmatrix} I^1_{q^1_o q^1_o} & & M_{q^1_o q_{t12}} & & & & M_{q^1_o q_{t0}} \\ & I^2_{q^2_o q^2_o} & M_{q^2_o q_{t12}} & & & & M_{q^2_o q_{t0}} \\ & & I^{12}_{q_{t12} q_{t12}} & & & & M_{q_{t12} q_{t0}} \\ & & & I^3_{q^3_o q^3_o} & & M_{q^3_o q_{t34}} & M_{q^3_o q_{t0}} \\ & sym & & & I^4_{q^4_o q^4_o} & M_{q^4_o q_{t34}} & M_{q^4_o q_{t0}} \\ & & & & & I^{34}_{q_{t34} q_{t34}} & M_{q_{t34} q_{t0}} \\ & & & & & & I^0_{q_{t0} q_{t0}} \end{bmatrix}.$$

The interior generalized masses $I^i_{q^i_o q^i_o}$ and stiffnesses $\Lambda^i_{q^i_o q^i_o}$ are the same as in Section 11.4. The boundary components generalized masses are

$$I^j_{q^i_t q^i_t} = \Phi^T_{t q^i_t} M^i_{tt} \Phi_{t q^i_t}.$$

Similarly the stiffnesses:

$$\Lambda^j_{q^i_t q^i_t} = \Phi^T_{t q^i_t} K^i_{tt} \Phi_{t q^i_t}.$$

The component-to-boundary mass and stiffness coupling terms are

$$M_{q^i_o q^j_t} = \Phi^{i,T}_{o q^i_o} M^i_{ot} \Phi_{t q^j_t}$$

and

$$K_{q^i_o q^j_t} = \Phi^{i,T}_{o q^i_o} K^i_{ot} \Phi_{t q^j_t},$$

where j is either $12, 34$ or 0. The boundary-to-boundary mass and stiffness coupling terms are

$$M_{q^j_t q^0_t} = \Phi^T_{t q^j_t} M^j_{tt} \Phi_{t q^0_t}$$

and

$$K_{q^j_t q^0_t} = \Phi^T_{t q^j_t} K^j_{tt} \Phi_{t q^0_t},$$

where j is either 12 or 34. This problem is of the same size as the number of columns in S, i.e. (d), and it has a structure of a main diagonal with some coupling matrices.

The following chart summarizes the effect of the dynamic reduction. The K_{dd} partition again is not a straightforward partition, it has been modified.

$$\begin{bmatrix} K_{aa} & \\ & [\,K_{dd}\,] \end{bmatrix}.$$

11.6   Multi-body analysis application

As an application example for dynamic reduction, let us consider the case in engineering analysis when the rigid components are not eliminated from the system as shown in Chapter 4. This happens for example when the rigid components have large displacements and they act as an independent, albeit attached body of the system, hence the name multi-body analysis. An example of such application is the steering mechanism, such as shown in Figure 11.1, of a car in relation to the car body.

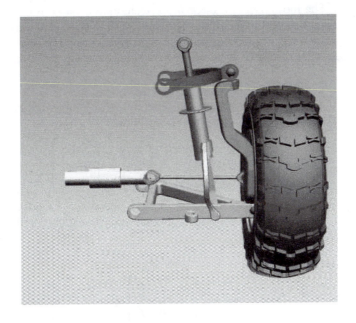

FIGURE 11.1 Steering mechanism

The relative motion between the flexible car body and the steering mechanism is described via a set of constraints dependent on the type of the joints used between the bodies. The rigid body system's (the steering mechanism) governing equation using Lagrange multipliers is

$$\begin{bmatrix} M & R^T \\ R & 0 \end{bmatrix} \begin{bmatrix} \ddot{q} \\ \lambda \end{bmatrix} = \begin{bmatrix} p \\ \mu \end{bmatrix}.$$

In this equation R is the constraint matrix and λ, μ are the Lagrange multipliers and the enforced accelerations, respectively. The physical meaning of the Lagrange multipliers is the reaction forces at the constraints. The q is the generalized displacement of the rigid body and $\ddot{q}$ is the acceleration. Note, that the governing equation does not contain stiffness terms for the rigid body system.

The actual structure of R is similar to that of R_{mg} introduced in Chapter 4. It is the mathematical representation of mechanical constraints between components of a multi-body system. An example is a simple hinge connecting two bodies, where the two bodies rotate freely about the same axis. The more difficult joints may have more difficult representation [5].

Most of the time rigid body components are an integrated part of a flexible structure. The most common integration technique is based on applying the dynamic reduction technique to the flexible partition of the structure. In other words, reduce the car body to a number of attachment points that specify the connection with the steering mechanism. This reduction process is detailed next.

Let us partition a finite element problem into the flexible or elastic (e partition) and rigid (r partition) parts [3] as follows:

$$
\begin{bmatrix} M_{rr} & M_{re} & R_r^T \\ M_{er} & M_{ee} & R_e^T \\ R_r & R_e & 0 \end{bmatrix} \begin{bmatrix} \ddot{q} \\ \ddot{u}_e \\ \lambda \end{bmatrix} = \begin{bmatrix} P_r \\ P_e \\ \mu \end{bmatrix} - \begin{bmatrix} 0 \\ K_{ee}u_e \\ 0 \end{bmatrix}.
$$

The M_{re} matrix represents the mass coupling between the flexible and rigid partitions. Note the presence of the K_{ee} term to represent the flexibility of the structure. The dynamic reduction transformation matrix for this partitioning is of form:

$$
T = \begin{bmatrix} I_r & 0 & 0 \\ 0 & \Phi_{ee_q} & 0 \\ 0 & 0 & I_r \end{bmatrix},
$$

where Φ_{ee_q} are the eigenvectors of the flexible part of the integrated structure. The number of eigenvectors found is e_q. They are the solution of the problem

$$
K_{ee}\Phi_{ee_q} = M_{ee}\Phi_{ee_q}\Lambda_{e_q e_q}.
$$

Let the eigenvectors be mass-orthogonal as

$$
I_{e_q e_q} = \Phi^t_{ee_q} M_{ee}\Phi_{ee_q}.
$$

The generalized stiffness is

$$
\Lambda_{e_q e_q} = \Phi^t_{ee_q} K_{ee}\Phi_{ee_q}.
$$

Let us introduce

$$
u_e = \Phi_{ee_q} u_{eq},
$$

and

$$
\ddot{u}_e = \Phi_{ee_q} \ddot{u}_{eq}.
$$

Pre-multiplying the above finite element problem by T^T and substituting yields

$$
\begin{bmatrix} M_{rr} & M_{re}\Phi_{ee_q} & R_r^T \\ \Phi^T_{ee_q}M_{er} & I_{e_q e_q} & \Phi^T_{ee_q}R_e^T \\ R_r & R_e\Phi_{ee_q} & 0 \end{bmatrix} \begin{bmatrix} \ddot{q} \\ \ddot{u}_{eq} \\ \lambda \end{bmatrix} \begin{bmatrix} P_r \\ \Phi^T_{ee_q}P_{eq} \\ \mu \end{bmatrix} - \begin{bmatrix} 0 \\ \Lambda_{e_q e_q}u_{eq} \\ 0 \end{bmatrix}.
$$

The dynamically reduced form (size $2r+e_q$ as opposed to $2r+e$ where $e >> e_q$) of the integrated equation of motion is now easier to solve. This technique is especially advantageous when the dynamic behavior of the structure is analyzed in a longer time interval and many time step solutions are required. The issue of the time integration is addressed in Chapter 14.

This formulation governs the rigid body system's behavior with the flexible body effects being presented with the reduction and preserved in the reduced form. It is also practical to take dynamic forces computed from this simulation back to the detailed flexible body for further analysis and detailed stress and response calculations.

Finally, it is prudent to point out that the dynamic reduction process described in this chapter needs to be generalized when the stiffness matrix is unsymmetric or complex, in which case the normal modes basis cannot be computed. In such case the quadratic eigenvalue problem presented in the last chapter is solved and the distinct left and right hand eigenvectors form the basis of reduction.

Naturally, the simple generalized mass and stiffness formulations do not hold anymore and they are replaced by the orthogonality criteria reflecting the distinct eigenvectors. Otherwise the concepts of the real dynamic reduction apply equally well. This, complex dynamic reduction technique is beyond our scope and will not be further discussed here.

References

[1] Chatelin, F.; Eigenvalues of matrices, Wiley, New York, 1993

[2] Hurty, W. C.; Dynamic analysis of structural systems using component modes, AIAA Journal, Vol. 3, No. 4, pp. 678-685, 1965

[3] Komzsik, L.; Integrated multi-body system and finite element analysis, Proceedings of 4th International Conference on Tools and Methods of Competitive Engineering, Wuhan, China, 2002

[4] Masson, G. et al; Parameterized reduced models for efficient optimization of structural dynamic behavior, AIAA-2002-1392, AIAA, 2002

[5] Nikravesh, P. E.; Computer-aided analysis of mechanical systems, Prentice Hall, New Jersey, 1988

[6] Ortega, J. M.; Matrix theory, Plenum Press, New York, 1987

[7] Saad, Y.; Numerical methods for large eigenvalue problems, Halsted Press, 1992

[8] Sleijpen, G. L. G., Eshof, J. V. D. and Smit, P.; Optimal *a priori* error-bounds for the Rayleigh-Ritz method, Math. Comp, Vol. 72, pp. 667-684, 2002

[9] Wamsler, M.; Retaining the influence of crucial local effects in mixed Guyan and modal reduction, Engineering with Computers, Vol 20, pp. 363-371, 2005

12

Component Mode Synthesis

The component mode synthesis procedure is a successive application of the static condensation and dynamic reduction and as such combines the techniques of Chapters, 8 and 11. It is applicable to any dynamic analysis problem, however, we continue to use the undamped, free vibration or normal modes problem of structural analysis as the basis for discussion of this method. See [1], [6] and [5] for early references.

This technique is also commonly executed with single and multiple components as well as on multiple levels. First we review the concept with the single-level case.

12.1 Single-level, single-component modal synthesis

The combined static condensation and dynamic reduction transformation matrix may be written as

$$R = TS,$$

where the static condensation transformation is

$$T = \begin{bmatrix} I_{oo} & G_{ot} \\ 0 & I_{tt} \end{bmatrix},$$

with the static condensation matrix $G_{ot} = -K_{oo}^{-1} K_{ot}$. The dynamic reduction matrix is

$$S = \begin{bmatrix} \Phi_{oq_o} & \\ & \Phi_{tq_t} \end{bmatrix}.$$

The modal space of the interior and the boundary are computed from

$$K_{oo}\Phi_{oq_o} = M_{oo}\Phi_{oq_o}\Lambda_{q_oq_o},$$

and

$$K_{tt}\Phi_{tq_t} = M_{tt}\Phi_{tq_t}\Lambda_{q_tq_t}.$$

Pre-multiplying the

$$(K_{aa} - \lambda M_{aa})u_a = 0$$

equation by R^T and introducing $R\overline{u}_d = u_a$ results in

$$R^T(K_{aa} - \lambda M_{aa})R\overline{u}_d = 0,$$

or

$$(\overline{K}_{dd} - \lambda \overline{M}_{dd})\overline{u}_d = 0,$$

where the notation $\overline{K}_{dd}$ reflects the fact that this matrix is now the result of both static condensation and dynamic reduction. In partitioned form

$$\begin{bmatrix} \Lambda_{q_o q_o} - \lambda I_{q_o q_o} & -\lambda M_{q_o q_t} \\ -\lambda M_{q_o q_t}^T & \Lambda_{q_t q_t} - \lambda I_{q_t q_t} \end{bmatrix} \begin{bmatrix} u_{q_o} \\ u_{q_t} \end{bmatrix} = 0.$$

Here the transformed mass coupling matrix is

$$M_{q_o q_t} = \Phi_{oq_o}^T \overline{M}_{ot} \Phi_{tq_t}$$

and due to the static condensation the stiffness matrix is uncoupled, while the mass is not, represented by $\overline{M}_{ot}$.

As before, due to the mass-orthogonality of the eigenvectors

$$\Phi_{oq_o}^T M_{oo} \Phi_{oq_o} = I_{q_o q_o},$$

and

$$\Phi_{tq_t}^T M_{tt} \Phi_{tq_t} = I_{q_t q_t}.$$

The transformed eigenvector components are defined by

$$\Phi_{oq_o} u_{q_o} = u_o$$

and

$$\Phi_{tq_t} u_{q_t} = u_t.$$

Note, that while the static condensation step is computationally exact, the dynamic reduction is not and therefore, the modal synthesis also contains truncation error. This is the topic of more discussion in Chapter 13.

Before we invoke an example let us reconcile the method shown here with the well-known Craig-Bampton method [2]. The R transformation matrix in our case, assuming full modal space, is

$$R = TS = \begin{bmatrix} \Phi_{oo} & G_{ot}\Phi_{tt} \\ 0 & \Phi_{tt} \end{bmatrix}.$$

The Craig-Bampton method's transformation matrix is

$$R_{CB} = \begin{bmatrix} \Phi_{oo} & G_{ot} \\ 0 & I_{tt} \end{bmatrix}.$$

It is easy to see that this form is the product of $R_{CB} = TS_{CB}$, where

$$S_{CB} = \begin{bmatrix} \Phi_{oo} \\ 0 & I_{tt} \end{bmatrix}.$$

This means that the Craig-Bampton method is a component mode synthesis method that does not do dynamic reduction on the boundary component. This is the only difference.

12.2 Mixed boundary component mode reduction

The method shown in the prior section has assumed a fixed boundary when computing the component modes. In some practical applications that is not appropriate. For example, when some of the boundary components are excited by loads, the dynamic responses computed from fixed boundary component mode reduction are poor [7]. The subject of this section is to discuss the case when some parts of the boundary are considered to be free and some part is fixed during the component mode computations, hence the name.

The matrix partitioning for the mixed boundary reduction is

$$K_{aa} = \begin{bmatrix} K_{cc} & & K_{c0} \\ 0 & K_{bb} & K_{bo} \\ K_{oc} & K_{ob} & K_{oo} \end{bmatrix} = \begin{bmatrix} \overline{K}_{tt} & K_{to} \\ K_{ot} & K_{oo} \end{bmatrix}.$$

Here the c partition is the boundary segment that is free. The b partition is the fixed boundary segment as before. Similar partitioning is used for the mass matrix.

$$M_{aa} = \begin{bmatrix} M_{cc} & & M_{c0} \\ 0 & M_{bb} & M_{bo} \\ M_{oc} & M_{ob} & M_{oo} \end{bmatrix} = \begin{bmatrix} \overline{M}_{tt} & M_{to} \\ M_{ot} & M_{oo} \end{bmatrix}.$$

Note, that in above partitioning the boundary was ordered first, followed by the interior, opposite to the order of the earlier sections. This is to reflect our focus on the split boundary and ease the notation, otherwise this issue is of no computational consequence. The two partitions are really interspersed after all.

Two static condensation matrices are computed for both boundary segments as

$$\begin{bmatrix} 0 & 0 & 0 \\ 0 & K_{bb} & K_{bo} \\ 0 & K_{ob} & K_{oo} \end{bmatrix} \begin{bmatrix} 0 \\ I_{bb} \\ G_{ob} \end{bmatrix} = \begin{bmatrix} 0 \\ P_b \\ 0 \end{bmatrix} \rightarrow G_{ob} = -K_{oo}^{-1} K_{ob},$$

and

$$\begin{bmatrix} K_{cc} & 0 & K_{c0} \\ 0 & 0 & 0 \\ K_{oc} & 0 & K_{oo} \end{bmatrix} \begin{bmatrix} I_{cc} \\ 0 \\ G_{oc} \end{bmatrix} = \begin{bmatrix} P_c \\ 0 \\ 0 \end{bmatrix} \rightarrow G_{oc} = -K_{oo}^{-1} K_{oc}.$$

To calculate the dynamic reduction mode shapes an eigenvalue problem of size $v = o + c$ is solved.

$$K_{vv} \Phi_{vz} = M_{vv} \Phi_{vz} \Lambda_{zz},$$

where z is the number of eigenpairs extracted. These mode shapes, however, are not directly used in the dynamic reduction. They are first partitioned as

$$\Phi_{vz} = \begin{bmatrix} \Phi_{oz} \\ \Phi_{cz} \end{bmatrix},$$

where the eigenvector partitions correspond to the interior and the free boundary, respectively. The constrained modes corresponding to static condensation are

$$\Phi_{oz}^1 = \Phi_{oz} - G_{oc} \Phi_{cz}.$$

The possible zero modes are partitioned out as

$$\Phi_{oz}^1 = \begin{bmatrix} \Phi_{oz}^0 & \Phi_{oz}^2 \end{bmatrix}$$

to compute the generalized mass

$$M_{zz}^2 = \Phi_{oz}^{2,T} M_{oo} \Phi_{oz}^2.$$

By introducing a scaling factor

$$S_{zz} = \sqrt{diag(M_{zz}^2)},$$

the modes are scaled as

$$\Phi_{oz}^3 = \Phi_{oz}^2 S_{zz}.$$

The superscripts represent various modifications to the vectors and matrices. The scaled modes produce a scaled generalized mass

$$M_{zz} = S_{zz}^T M_{zz}^2 S_{zz}.$$

To verify the non-singularity of this matrix, the generalized mass is factored

$$M_{zz} = L_{zz}^T D_{zz} L_{zz}.$$

Dependent vectors are purged from the eigenvectors if

$$M_{zz}(ii)/D_{zz}(i) > \epsilon_{dep},$$

where the ϵ_{dep} is a dependency cut-off level. With these the generalized stiffness matrix is

$$K_{zz} = \Phi_{oz}^{3,T} K_{oo} \Phi_{oz}^3.$$

Finally, the scaled generalized modes are computed from

$$K_{zz}\Phi_{oz}^4 = M_{zz}\Phi_{oz}^4 \Lambda_{zz}.$$

These modes are going to be used for the dynamic reduction part. In specific, the reduction matrix for the mixed boundary case will be

$$R = \begin{bmatrix} G_{ob} & G_{oc} & \Phi_{oz}^4 \\ I_{bb} & 0 & 0 \\ 0 & I_{cc} & 0 \end{bmatrix} = \begin{bmatrix} G_{ot} & \Phi_{oz}^4 \\ I_{tt} & 0 \end{bmatrix}$$

$$G_{ot} = \begin{bmatrix} G_{ob} & G_{oc} \end{bmatrix}.$$

At this point the two methods are reconciled, albeit with a different transformation matrix. The reduced stiffness matrix is

$$\overline{K}_{dd} = R^T K_{aa} R = \begin{bmatrix} K_{tt} & 0 \\ 0 & K_{qq} \end{bmatrix},$$

with

$$K_{tt} = \overline{K}_{tt} + K_{to} G_{ot}$$

and

$$K_{qq} = \Phi_{oz}^{T,4} K_{oo} \Phi_{oz}^4.$$

The mass reduction, due to the boundary-coupling terms, is a bit more difficult but conceptually similar.

$$\overline{M}_{dd} = R^T M_{aa} R = \begin{bmatrix} M_{tt} & M_{tq} \\ M_{qt} & M_{qq} \end{bmatrix},$$

with

$$M_{tt} = \overline{M}_{tt} + M_{to}G_{ot} + G_{ot}^T M_{ot} + G_{ot}^T M_{oo} G_{ot},$$

$$M_{qt} = \Phi_{oz}^{T,4} M_{ot} + \Phi_{oz}^{T,4} M_{oo} G_{ot},$$

and

$$M_{qq} = \Phi_{oz}^{T,4} M_{oo} \Phi_{oz}^4.$$

The full-size result is obtained from

$$u_a = R u_d.$$

In detailed partitions this equation is:

$$\begin{bmatrix} u_t \\ u_o \end{bmatrix} = \begin{bmatrix} I_{tt} & 0 \\ G_{ot} & G_{oq} \end{bmatrix} \begin{bmatrix} u_t \\ u_q \end{bmatrix}.$$

The first column partition of

$$\begin{bmatrix} I_{tt} \\ G_{ot} \end{bmatrix}$$

represents the "constraint modes" of the component due to (and including the) unit boundary motion.

The second partition of

$$\begin{bmatrix} 0 \\ G_{oq} \end{bmatrix}$$

represents the "fixed boundary modes" of the component assuming (and including the) zero boundary motion.

In contrast, for the fixed method all of the boundary has zero motion during the mode shape computation, since

$$G_{oq} = \Phi_{oz}.$$

For the mixed boundary method only part of the boundary has zero motion during the mode shape computation, since

$$G_{oq} = \begin{bmatrix} \Phi_{oz}^3 & \Phi_{zz} \end{bmatrix}.$$

In the following sections of this chapter, we will assume the fixed boundary approach for simplicity, however, the mixed boundary reduction results are applicable as well.

12.3   Computational example

Let us now clarify the modal synthesis procedure with the help of a simple numerical example of the eigenvalue problem. Consider the generalized eigenvalue problem with the stiffness matrix used in Section 8.2.

$$K_{aa} = \begin{bmatrix} 2 & 1 & 0 \\ 1 & 3 & 1 \\ 0 & 1 & 4 \end{bmatrix}.$$

We will use the mass matrix of

$$M_{aa} = \begin{bmatrix} 1 & 1/2 & 0 \\ 1/2 & 1 & 1/2 \\ 0 & 1/2 & 1 \end{bmatrix}.$$

The analytic global eigenvalue solution for the

$$(K_{aa} - \lambda M_{aa})u_a = 0$$

eigenvalue problem is

$$\Lambda = \begin{bmatrix} 2 & 0 & 0 \\ 0 & 3 & 0 \\ 0 & 0 & 6 \end{bmatrix},$$

and

$$U = \begin{bmatrix} 1 & -1/2 & -1/3 \\ 0 & 1 & 2/3 \\ 0 & 1/2 & -2/3 \end{bmatrix}.$$

Let us again partition the problem into the interior consisting of the principal 2 by 2 minor

$$K_{oo} = \begin{bmatrix} 2 & 1 \\ 1 & 3 \end{bmatrix}$$

and

$$M_{oo} = \begin{bmatrix} 1 & 1/2 \\ 1/2 & 1 \end{bmatrix},$$

as well as

$$K_{ot} = \begin{bmatrix} 0 \\ 1 \end{bmatrix}$$

and

$$M_{ot} = \begin{bmatrix} 0 \\ 1/2 \end{bmatrix}.$$

The boundary components are

$$K_{tt} = \begin{bmatrix} 4 \end{bmatrix}$$

and

$$M_{tt} = \begin{bmatrix} 1 \end{bmatrix}.$$

For this stiffness matrix we have calculated the static condensation matrix in Chapter 8 as

$$G_{ot} = \begin{bmatrix} 1/5 \\ -2/5 \end{bmatrix}.$$

The mass coupling matrix with this is

$$\overline{M}_{ot} = \begin{bmatrix} 0 \\ 1/2 \end{bmatrix} + \begin{bmatrix} 1 & 1/2 \\ 1/2 & 1 \end{bmatrix} \begin{bmatrix} 1/5 \\ -2/5 \end{bmatrix} = \begin{bmatrix} 0 \\ 1/5 \end{bmatrix}.$$

The Schur complement result of the static condensation was

$$\overline{K}_{tt} = \begin{bmatrix} 18/5 \end{bmatrix}.$$

The corresponding mass complement:

$$\overline{M}_{tt} = \begin{bmatrix} 1 \end{bmatrix} + \begin{bmatrix} 0 & 1/2 \end{bmatrix} \begin{bmatrix} 1/5 \\ -2/5 \end{bmatrix} + \begin{bmatrix} 1/5 & -2/5 \end{bmatrix} \begin{bmatrix} 0 \\ 1/2 \end{bmatrix} +$$

$$\begin{bmatrix} 1/5 & -2/5 \end{bmatrix} \begin{bmatrix} 1 & 1/2 \\ 1/2 & 1 \end{bmatrix} \begin{bmatrix} 1/5 \\ -2/5 \end{bmatrix} = \begin{bmatrix} 18/25 \end{bmatrix}.$$

Finally, the matrices of the statically condensed eigenproblem are

$$\overline{K} = \begin{bmatrix} 2 & 1 & 0 \\ 1 & 3 & 0 \\ 0 & 0 & 18/5 \end{bmatrix}$$

and

$$\overline{M} = \begin{bmatrix} 1 & 1/2 & 0 \\ 1/2 & 1 & 1/5 \\ 0 & 1/5 & 18/25 \end{bmatrix}.$$

The next step is to execute the dynamic reduction of this problem. This is, however, the same matrix problem we have used in discussing the dynamic reduction in Section 11.3, so the solution of the component mode synthesis problem comes from that of the solution of the example in Chapter 11. The eigenvalues were

$$\Lambda = \begin{bmatrix} 2 & 0 & 0 \\ 0 & 3 & 0 \\ 0 & 0 & 6 \end{bmatrix},$$

which are identical to the original problem's solution. The eigenvectors (back-transformed with respect to the dynamic reduction) were:

$$\overline{U} = \begin{bmatrix} 1 & -1/2 & 3/10 \\ 0 & 1 & -3/5 \\ 0 & 5/12 & 1 \end{bmatrix}.$$

The global eigenvectors are obtained by redoing the static condensation also. For the interior:

$$U_o = \begin{bmatrix} 1 & -1/2 & 3/10 \\ 0 & 1 & -3/5 \end{bmatrix} + \begin{bmatrix} 1/5 \\ -2/5 \end{bmatrix} \begin{bmatrix} 0 & 5/12 & 1 \end{bmatrix} = \begin{bmatrix} 1 & -5/12 & 1/2 \\ 0 & 5/6 & -1 \end{bmatrix}.$$

For the boundary:

$$U_t = \begin{bmatrix} 0 & 5/12 & 1 \end{bmatrix}.$$

Hence, the final eigenvectors are

$$U = \begin{bmatrix} 1 & -5/12 & 1/2 \\ 0 & 5/6 & -1 \\ 0 & 5/12 & 1 \end{bmatrix},$$

which are the same as the analytic global solution apart from a scalar multiplier.

In the example, the full eigenspectrum of both the boundary and interior is obtained, $t = q_t$ and $o = q_o$, therefore, there was no reduction and the problem was merely transformed. Naturally, the results then are accurate. The smaller the number of eigenvectors compared to the rank of the matrix, the less accurate the component mode synthesis becomes due to the truncation error.

In the following we discuss the two stages of the component mode synthesis reduction technique for the multiple-component cases.

12.4 Single-level, multiple-component modal synthesis

The problem at hand with multiple components is partitioned as

$$\begin{bmatrix} K_{oo}^1 - \lambda M_{oo}^1 & & & K_{ot}^1 - \lambda M_{ot}^1 \\ & \ddots & & \vdots \\ & & K_{oo}^i - \lambda M_{oo}^i & K_{ot}^i - \lambda M_{ot}^i \\ & & & \vdots \\ K_{ot}^{T,1} - \lambda M_{ot}^{T,1} & \cdots & K_{ot}^{T,i} - \lambda M_{ot}^{T,i} & \cdots & K_{tt} - \lambda M_{tt} \end{bmatrix} \begin{bmatrix} u_o^1 \\ \vdots \\ u_o^i \\ \vdots \\ u_t \end{bmatrix} = 0.$$

For each component the static reduction to the boundary is facilitated by

$$G_{ot}^i = -K_{oo}^{-1,i} K_{ot}$$

where i goes from 1 to n and the computations are independent. The simultaneous static reduction for all components is formally executed by the

$$T = \begin{bmatrix} I_{oo}^1 & & & G_{ot}^1 \\ & \ddots & & \vdots \\ & & I_{oo}^i & G_{ot}^i \\ & & & \ddots & \vdots \\ & & & & I_{tt} \end{bmatrix}$$

transformation matrix. Pre-multiplying with T^T and introducing $T\bar{u}_a = u_a$ as

$$T^T[K_{aa} - \lambda M_{aa}]T\bar{u}_a = 0$$

results in

$$[\overline{K}_{aa} - \lambda \overline{M}_{aa}]\bar{u}_a = 0.$$

In detail we get

$$\begin{bmatrix} K_{oo}^1 - \lambda M_{oo}^1 & & & -\lambda \overline{M}_{ot}^1 \\ & \ddots & & \vdots \\ & & K_{oo}^i - \lambda M_{oo}^i & -\lambda \overline{M}_{ot}^i \\ & & & \ddots & \vdots \\ -\lambda \overline{M}_{ot}^{T,1} & \cdots & -\lambda \overline{M}_{ot}^{T,i} & \cdots & \overline{K}_{tt} - \lambda \overline{M}_{tt} \end{bmatrix} \begin{bmatrix} \bar{u}_o^1 \\ \vdots \\ \bar{u}_o^i \\ \vdots \\ u_t \end{bmatrix} = 0,$$

where

$$\overline{K}_{tt} = K_{tt} + \Sigma_{i=1}^n K_{ot}^{T,i} G_{ot}^i,$$

$$\overline{M}_{tt} = M_{tt} + \Sigma_{i=1}^n (M_{ot}^{T,i} G_{ot}^i + G_{ot}^{T,i} M_{ot}^i + G_{ot}^{T,i} M_{oo}^i G_{ot}^i),$$

and the mass coupling term is

$$\overline{M}_{ot}^i = M_{ot}^i + M_{oo}^i G_{ot}^i.$$

Note, that the coupling stiffness term has been eliminated and the eigenvalues are invariant under this congruence transformation. The eigenvector interior components are transformed as

$$\bar{u}_o^i = u_0^i - G_{ot}^i u_t,$$

which will be the basis for result recovery. Note, that the u_t term is unaffected during the transformation by T.

To complete the component mode synthesis we now apply dynamic reduction to the statically condensed problem. The dynamic behavior of the components is represented by the fixed boundary component modes

$$K_{oo}^i \Phi_{oq_o^i}^i = M_{oo}^i \Phi_{oq_o^i}^i \Lambda_{o_q^i q^i}^i$$

and the boundary modes of

$$\overline{K}_{tt} \Phi_{tq_t} = \overline{M}_{tt} \Phi_{tq_t} \Lambda_{q_t q_t}.$$

Note, that the boundary modes are based on the statically condensed boundary problem and the number of mode shapes found for each component may be different.

The dynamic reduction transformation matrix in this form is

$$
S = \begin{bmatrix} \Phi^1_{oq_o^1} & & & \\ & \ddots & & \\ & & \Phi^i_{oq_o^i} & \\ & & & \ddots \\ & & & & \Phi_{tq_t} \end{bmatrix}.
$$

Observe that this matrix is rectangular as the number of mode shapes extracted from each component is less than the size of the component. S has $d = q_t + \Sigma^n_{i=1} q_o^i$ number of columns and $t + \Sigma^n_{i=1} o^i$ number of rows. Appropriate multiplication of the equilibrium equation with S^T and the introduction of $S\overline{u}_d = \overline{u}_a$ as

$$
S^T(\overline{K}_{aa} - \lambda\overline{M}_{aa})S\overline{u}_d = 0
$$

yields the modal synthesis reduced form of

$$
(\overline{K}_{dd} - \lambda\overline{M}_{dd})\overline{u}_d = 0.
$$

In detail the reduced form of the modal synthesis is

$$
\begin{bmatrix} \Lambda^1_{q_o^1 q_o^1} - \lambda I^1_{q_o^1 q_o^1} & & & -\lambda M^1_{q_o^1 q_t} \\ & \ddots & & \vdots \\ & & \Lambda^i_{q_o^i q_o^i} - \lambda I^i_{q_o^i q_o^i} & -\lambda M^i_{q_o^i q_t} \\ & & & \ddots \\ -\lambda M^{T,1}_{q_t q_o^1} & \cdots & -\lambda M^{T,i}_{q_t q_o^i} & \Lambda_{q_t q_t} - \lambda I_{q_t q_t} \end{bmatrix} \begin{bmatrix} u^1_q \\ \vdots \\ u^i_q \\ \vdots \\ \overline{u}_{q_t} \end{bmatrix} = 0.
$$

Here the generalized stiffnesses are

$$
\Lambda^i_{q_o^i q_o^i} = \Phi^{T,i}_{oq_o^i} K^i_{oo} \Phi^i_{oq_o^i}
$$

and

$$
\Lambda_{q_t q_t} = \Phi^T_{tq} \overline{K}_{tt} \Phi_{tq}.
$$

The generalized mass matrices are

$$
I^i_{q_o^i q_o^i} = \Phi^{T,i}_{oq_o^i} M^i_{oo} \Phi^i_{oq_o^i}
$$

and

$$
I_{q_t q_t} = \Phi^T_{tq_t} \overline{M}_{tt} \Phi_{tq_t},
$$

since the eigenvectors are again mass-orthogonal. The mass coupling term is

$$
M^i_{q_o^i q_t} = \Phi^{T,i}_{oq_o^i} \overline{M}^i_{ot} \Phi_{tq_t}.
$$

The eigenvalues are again invariant under this congruence transformation. The eigenvector components are transformed as

$$
\Phi^i_{oq_o^i} u^i_q = \overline{u}^i_o
$$

and

$$\Phi_{tq_t} u_{q_t} = u_t.$$

The reduced eigenvalue problem contains matrices with very specific structure. Namely, the stiffness matrix is diagonal, but the mass is not (although it is very sparse). This gives the idea, sometimes used in commercial analyses, to solve the mode synthesized problem in inverse form, This method of replacing the mass and stiffness matrices allows for an efficient Lanczos method solution, where the mass matrix needs to be used in every step of the operation. The shortcoming of this computationally speedy idea is the need for special handling of the possible rigid body modes remaining in the reduced problem. These modes become computationally infinite in the inverted solution scheme and as such will not appear in the solution, however, they are of particular interest of the engineer.

The reduced eigenvalue problem is again of order $d = q_t + \Sigma_{i=1}^n q_o^i$ size as opposed to the original order $t + \Sigma_{i=1}^n o^i$ size.

The eigenvectors of the reduced problem are twice transformed. Therefore, the final eigenvectors will be recovered in two steps. First, the dynamic reduction is recovered as

$$\overline{u}_o^i = \Phi_{oq_o^i}^{T,i} u_q^i$$

and

$$u_t = \Phi_{tq_t} \overline{u}_t.$$

Secondly, the effects of the static reduction are accounted for as

$$u_o^i = \overline{u}_o^i + G_{ot}^i u_t.$$

Remember, u_t does not change during static condensation.

12.5 Multiple-level modal synthesis

The combined static condensation and dynamic reduction transformation matrix may be written as

$$R = TS.$$

The T and S matrices were described in Sections 8.4 and 11.5, respectively. Pre-multiplying

$$[K_{aa} - \lambda M_{aa}] u_a = 0$$

by R^T and substituting

$$R\overline{u}_d = u_a$$

yields

$$R^T[K_{aa} - \lambda M_{aa}]R\overline{u}_d = 0.$$

Executing the multiplications result in

$$[\overline{K}_{dd} - \lambda \overline{M}_{dd}]\overline{u}_d = 0.$$

The detailed structure of the reduced stiffness matrix is

$$\overline{K}_{dd} = \begin{bmatrix} \Lambda^1_{q^1_o q^1_o} & & & & & & \\ & \Lambda^2_{q^2_o q^2_o} & & & & & \\ & & \overline{\Lambda}^{12}_{q_t 12 \, q_t 12} & & & & \\ & & & \Lambda^3_{q^3_o q^3_o} & & & \\ & & & & \Lambda^4_{q^4_o q^4_o} & & \\ & & & & & \overline{\Lambda}^{34}_{q_t 34 \, q_t 34} & \\ & & & & & & \overline{\Lambda}^0_{q_t 0 \, q_t 0} \end{bmatrix}.$$

The $\overline{\Lambda}$ notation reflects the fact that the boundary eigenvalue solution is computed from the statically condensed boundary problem. The detailed structure of the reduced mass matrix is

$$\overline{M}_{dd} = \begin{bmatrix} I^1_{q^1_o q^1_o} & & \overline{M}_{q^1_o q_t 12} & & & & \overline{M}_{q^1_o q_t 0} \\ & I^2_{q^2_o q^2_o} & \overline{M}_{q^2_o q_t 12} & & & & \overline{M}_{q^2_o q_t 0} \\ & & I^{12}_{q_t 12 \, q_t 12} & & & & \overline{M}_{q_t 12 \, q_t 0} \\ & & & I^3_{q^3_o q^3_o} & & \overline{M}_{q^3_o q_t 34} & \overline{M}_{q^3_o q_t 0} \\ & sym & & & I^4_{q^4_o q^4_o} & \overline{M}_{q^4_o q_t 34} & \overline{M}_{q^4_o q_t 0} \\ & & & & & I^{34}_{q_t 34 \, q_t 34} & \overline{M}_{q_t 34 \, q_t 0} \\ & & & & & & I^0_{q_t 0 \, q_t 0} \end{bmatrix}.$$

The $\overline{M}$ notation represents the fact that the mass coupling matrices underwent static condensation prior to the dynamic reduction. All generalized masses and stiffnesses are the same as in Sections 8.4 and 11.4. This problem is of the same size as the reduced problem after multilevel dynamic reduction, however, now the stiffness matrix is diagonal. This fact provides further computational advantages.

The following is the summary chart for this stage of the reduction.

$$\begin{bmatrix} K_{aa} & \\ & \begin{bmatrix} \overline{K}_{tt} & \\ & [\overline{K}_{dd}] \end{bmatrix} \end{bmatrix}.$$

Here again, $\overline{K}_{tt}$ and $\overline{K}_{dd}$ are not direct partitions.

12.6   Component mode synthesis case study

The effectiveness of the computational reduction techniques is viewed through a case study. This model was also selected from the automobile industry, although similar characteristics may also be found in the aerospace industry, specifically in airplane fuselage models. The example was similar to the convertible car body as shown in Figure 12.1.

FIGURE 12.1 Convertible car body

The model incorporated a variety of structural components containing various element types as shown in Table 12.1.

The rigid elements included rigid bar elements, such as the one used in the mechanical example in Chapter 3. They also included various forms of multipoint constraints which are used regularly in the auto industry to model spot welds. As the large number of these indicate, there are quite a few of these

TABLE 12.1

Element types of case study automobile model

Element type	Number of elements
4 noded quadrilateral	274,192
3 noded triangular	59,729
8 noded hexahedral	3,645
4 noded tetrahedral	532,720
Rigid elements	12,830

in a car body.

The model had 1,301,381 nodes and the statistics of the degrees of freedom are presented in Table 12.2.

TABLE 12.2

Problem statistics

Partition	Size
Global degrees of freedom = g	7,808,286
multi-point constraints = m	170,345
single-point constraints = s	3,101,117
Free set size = f	4,536,824

The rather large number of single point constraints is in part due to boundary conditions. Most of them, however, are related to the three dimensional elements and were found by the automatic elimination process discussed in Chapter 5.

Various scenarios of component mode synthesis were executed on this model to obtain all the eigenmodes up to 600 Hz. Since the problem was extremely compute intensive, establishing a serial base line was impractical. Instead, a base line with 64 processor execution of the hierarchic parallel normal modes analysis, demonstrated in Section 9.6 was found. The comparison statistics are in Table 12.3.

In all three executions a 64 node Linux cluster with dual core 1.85 GHz processors, 4 Gigabytes of memory and 50 Gigabytes of disk per node was used. It is noticeable that the component mode synthesis finds less modes than the computationally exact baseline solution, a fact the engineer needs to be aware of. The discrepancy, of course, is due to the approximation involved

TABLE 12.3
Execution statistics

Components	Elapsed time min:sec	I/O amount Gigabytes	Number of modes
-	829:53	1,814	5261
256	201:01	707	5176
512	189:22	888	5132

in the dynamic reduction.

It appears that for this example 256 components was the optimal, considering that the incremental elapsed time improvement for 512 components was negligible. Assuming that the baseline run would be about 20 times faster than a serial run (an assumption justified by Figure 9.4) the speed-up of the 512 component run over the serial one would be approaching 90. Properly applied component mode synthesis normal modes analyses often produce two orders of magnitude (hundredfold) improvements over a serial execution. In many cases, the serial execution can not even be done, due to machine limitations.

Such a performance advantage of the component mode synthesis based analysis solutions is obviously of great industrial importance. In essence a multi-day computational job is reduced to an overnight execution. Due to the current industry tendency of ever widening frequency ranges of interest [4], the component modal synthesis approach seems to be most practical. That is of course with the understanding of the engineering approximations involved.

This concludes the second part of the book dealing with computational reduction techniques.

References

[1] Benfield, W. A. and Hruda, R. F.; Vibration analysis of structures by component mode substitution, AIAA Journal, Vol. 9, No. 7, pp. 1255-1261, 1971

[2] Craig, R. R., Jr. and Bampton, M. C. C.; Coupling of substructures for dynamic analysis, AIAA Journal, Vol. 6, No. 7, pp. 1313-1319, 1968

[3] Komzsik, L.; A comparison of Lanczos method and AMLS for eigen-

value analysis, US National Conference on Computational Mechanics, Structural Dynamics Mini-symposium, Albuquerque, 2003

[4] Kropp, A. and Heiserer, D.; Efficient broadband vibro-acoustic analysis of passenger car bodies using a FE-based component mode synthesis approach, Proc. of 5th World Congress on Computational Mechanics, Vienna, 2002

[5] MacNeal, R. H.; A hybrid method of component mode synthesis, Computers and Structures, Vol. 1, No. 4, pp. 389-412, 1971

[6] Rubin, S.; Improved component-mode representation for structural dynamic analysis, AIAA Journal, Vol. 12, No. 8, pp. 995-1006, 1975

[7] Wamsler, M., Komzsik, L., and Rose, T., Combination of quasi-static and truncated system mode shapes, Proceedings of NASTRAN European User's Conference, The MacNeal-Schwendler Corporation, Amsterdam, 1992

Part III

Engineering Solution Computations

13

Modal Solution Technique

We now embark on the road to calculate the engineering solutions, the topic of the third, final, part of the book. This chapter will discuss a frequently used solution technique for the analysis of the transient behavior of a mechanical system, the modal solution technique. This is the application of the computational reduction techniques to the forced vibration problem.

13.1   Modal solution

The subject of the modal solution is the efficient solution of the forced, damped vibration problem of:

$$[M_{aa}\ddot{v}_a(t) + B_{aa}\dot{v}_a(t) + K_{aa}v_a(t)] = F_a(t).$$

The modal solution is based on the free, undamped vibrations of the system, represented by a linear, generalized eigenvalue problem of

$$(K_{aa} - \lambda M_{aa})\phi_a = 0,$$

where $\lambda = -\omega^2$. Let the matrix Φ_{ah} contain h eigenvectors. They may have been computed directly by solving the a partition eigenvalue problem, via dynamic reduction or component mode synthesis.

The essence of modal solution is to introduce the modal displacement w, defined as

$$v_a(t) = \Phi_{ah}\ w_h(t).$$

The length of the modal displacement vector is h. This equation actually will be used to transform the modal solution $w_h(t)$ to the original $v_a(t)$ solution of the engineering problem.

Similarly, modal velocities

$$\dot{v}_a(t) = \Phi_{ah}\dot{w}_h(t),$$

and accelerations

$$\ddot{v}_a(t) = \Phi_{ah}\ddot{w}_h(t)$$

are introduced. Pre-multiplying the forced, damped vibration equilibrium equation by Φ_{ah}^T

$$\Phi_{ah}^T[M_{aa}\ddot{v}_a(t) + B_{aa}\dot{v}_a(t) + K_{aa}v_a(t)] = \Phi_{ah}^T F_a(t),$$

and substituting the modal quantities, called modal coordinates we get

$$\Phi_{ah}^T M_{aa}\Phi_{ah}\ddot{w}_h(t) + \Phi_{ah}^T B_{aa}\Phi_{ah}\dot{w}_h(t) + \Phi_{ah}^T K_{aa}\Phi_{ah}w_h(t) = \Phi_{ah}^T F_a(t).$$

Let us define modal matrices as modal mass

$$M_{hh} = \Phi_{ah}^T M_{aa}\Phi_{ah},$$

modal damping

$$B_{hh} = \Phi_{ah}^T B_{aa}\Phi_{ah},$$

and modal stiffness

$$K_{hh} = \Phi_{ah}^T K_{aa}\Phi_{ah}.$$

The modal load is

$$F_h(t) = \Phi_{ah}^T F_a(t).$$

Substituting all of these results in the modal form of the equation of motion

$$M_{hh}\,\ddot{w}_h(t) + B_{hh}\,\dot{w}_h(t) + K_{hh}\,w_h(t) = F_h(t).$$

This equation is now of order h, much smaller than the original equation of motion, and as such is much cheaper and easier to solve. The time domain analysis technique shown in Chapter 14 is eminently applicable to these set of equations.

The summary chart now demonstrates the relationship between the analysis and the modal solution set as

$$\begin{bmatrix} K_{aa} & \\ & [K_{hh}] \end{bmatrix}.$$

In practical modal solution techniques the eigenvectors of the free vibration problem are mass normalized, so M_{hh} is the identity matrix and K_{hh} is a diagonal matrix, containing the eigenvalues.

The damping definition style can have a major effect on the efficiency of the modal solution, where the reduction results in diagonal stiffness and mass matrices in the modal basis, but the reduced damping matrix can become full, causing the cost of such analysis to grow greatly. In the midst of all the uncertainty about the actual physics, and the great growth in computational costs, the most common damping used in modal analysis is called modal damping.

An experienced engineer who has modeled and tested similar structures has some notion of what the modal damping for different types of modes is likely to be on the structure being analyzed. Based on this a modal damping value for each mode may be established, resulting in a diagonal modal damping matrix. This modal damping coefficient is multiplied by the natural frequency, resulting in the method called proportional damping. The modal equation in this case may be decoupled into a series of scalar equations as:

$$\ddot{w}_h[j](t) + 2\xi_j\omega_j\dot{w}_h[j](t) + \omega_j^2 w_h[j](t) = f_j(t) \quad j = 1, 2..h.$$

Here $\omega_j^2 = \lambda_j$ is the j-th eigenvalue and $w_h[j]$ is the j-th term in the modal solution vector. The modal damping matrix with proportional damping is defined as

$$B_{hh}[j, j] = 2\xi_j\omega_j,$$

and is a diagonal matrix. The j-th modal load is

$$f_j(t) = \Phi_{ah}[, j]^T F(t),$$

where $\Phi_{ah}[, j]$ is the j-th eigenvector (the j the column of Φ_{ah}).

13.2 Truncation error in modal solution

The truncation error is introduced by having less than the full eigenspectrum of the K_{aa}, M_{aa} matrix pencil represented in the reduced forms. This issue is also of paramount importance to the dynamic reduction techniques presented in the earlier chapters.

To assess the error accrued by the modal solution, let us concentrate on the undamped case of the modal problem and execute a Laplace transformation of form

$$w(\omega) = \int_0^\infty w(t)e^{-\omega t}dt.$$

Assuming zero initial conditions $w(0) = 0$ and $\dot{w}(0) = 0$ we get:

$$(\omega^2 + \lambda_j)\, w_h[j](\omega) = f_j(\omega),$$

for all $j = 1, 2, ..h$. From this the modal solution component is

$$w_h[j](\omega) = \frac{1}{\omega^2 + \lambda_j} f_j(\omega).$$

For simplicity in the following, the (ω) notation will be ignored as the frequency-dependence is obvious. Combining the scalar equations into a matrix equation again, we obtain:

$$w_h = \begin{bmatrix} \cdot & & \\ & \cdot & \\ & & \frac{1}{\omega^2 + \lambda_j} & \\ & & & \cdot \\ & & & & \cdot \end{bmatrix} \Phi_{ah}^T F,$$

where the right-hand side has been back-transformed to the original a-partition load vector. Back-transforming the modal solution results in

$$v_a = \Phi_{ah} w_h = Z^m F,$$

with the flexibility matrix:

$$Z^m = \Phi_{ah} \begin{bmatrix} \cdot & & \\ & \cdot & \\ & & \frac{1}{\omega^2 + \lambda_j} & \\ & & & \cdot \\ & & & & \cdot \end{bmatrix} \Phi_{ah}^T.$$

This solution is unfortunately not exact, since not all the eigenvectors of the matrix pencil are used, i.e. $h < a$. This is the cause of the truncation error.

Assume now that the complete eigenspace consists of the calculated portion Φ_{ah}^m and the truncated component of Φ_{ak}^r and $a = h + k$. Let us represent this by the partitioning

$$\Phi_{aa} = \begin{bmatrix} \Phi_{ah}^m & \Phi_{ak}^r \end{bmatrix}.$$

Taking this into consideration the exact solution is now

$$v_a = Z^m F_a + Z^r F_a,$$

where the two distinct flexibility matrices are called modal and residual flexibility matrices, respectively. Note, that the latter is not computed. The truncated component of the response solution is simply:

$$v_a^r = Z^r F_a = v_a - Z^m F_a.$$

As the exact solution v_a is not known, this formula is not useful for measuring the error. It is, however, useful in that it indicates how to reduce the truncation error. This is the topic of the next section.

13.3 The method of residual flexibility

The challenge in improving the accuracy of the modal solution is to account for the effect of the residual flexibility (Z^r), without computing the residual modes. Hence, the method described in the following is called the residual flexibility method [5].

Let us also assume that the eigenvectors contained in the residual set are related to eigenvalues well above the frequency of interest, the high frequency modes. Then $\lambda_j >> \omega^2$ and we may approximate the residual flexibility matrix as:

$$Z^r \approx \Phi_{ak}^r \begin{bmatrix} \cdot & & & & \\ & \cdot & & & \\ & & \frac{1}{\lambda_j} & & \\ & & & \cdot & \\ & & & & \cdot \end{bmatrix} \Phi_{ak}^{r,T}.$$

This suggests that the residual flexibility may be presented via static shapes. For this we consider the general case when the load consists of a time-dependent $H_a(t)$ and a time invariant G_a, sometimes called scaling, component as

$$F_a(t) = G_a H_a(t).$$

The static response ψ_a of the structure at any point in time is the solution of

$$K_{aa}\psi_a = G_a.$$

The complete modal representation of the stiffness matrix is

$$\Lambda_{aa} = \Phi_{aa}^T K_{aa} \Phi_{aa},$$

where

$$\Lambda_{aa} = \begin{bmatrix} \lambda_1 & & & & \\ & \cdot & & & \\ & & \cdot & & \\ & & & \lambda_j & \\ & & & & \cdot \\ & & & & & \lambda_a \end{bmatrix}.$$

Inverting both sides produces

$$\Lambda_{aa}^{-1} = \Phi_{aa}^{-1} K_{aa}^{-1} \Phi_{aa}^{-1,T}.$$

Pre- and post-multiplying both sides and using the orthogonality property of the eigenvectors results

$$K^{-1} = \Phi_{aa} \Lambda_{aa}^{-1} \Phi_{aa}^T.$$

Assuming the partitioning

$$\Phi_{aa} = \begin{bmatrix} \Phi_{ah}^m & \Phi_{ak}^r \end{bmatrix},$$

the inverse is of the form:

$$K_{aa}^{-1} = \Phi_{ah}^m \Lambda^{-1,m} \Phi_{ah}^{m,T} + \Phi_{ak}^r \Lambda^{-1,r} \Phi_{ak}^{r,T} = Z^m + Z^r.$$

Here $\Lambda^{-1,m}, \Lambda^{-1,r}$ are diagonal matrices containing the inverses of the eigenvalues corresponding to the computed modal space and the uncomputed residual space, respectively.

The residual flexibility from this equation is

$$Z^r = K_{aa}^{-1} - Z^m = K_{aa}^{-1} - \Phi_{ah}^m \Lambda^{-1,m} \Phi_{ah}^{m,T}.$$

The static solution component (residual vector) representing the residual flexibility is:

$$\psi_a^r = Z^r G_a = K_{aa}^{-1} G_a - \Phi_{ah}^m \Lambda^{-1,m} \Phi_{ah}^{m,T} G_a.$$

Note, that the right-hand side contains only computed quantities. Hence, the truncated component of the response solution is

$$v_a^r = \psi_a^r H_a(t).$$

Finally, the improved response solution corrected for the truncated residual flexibility is

$$v_a = \begin{bmatrix} \Phi_{ah}^m & \psi_a^r \end{bmatrix} \begin{bmatrix} w_h \\ H_a(t) \end{bmatrix}.$$

The disadvantage of this method is its computational expense. The stiffness matrix has to be factored and forward-backward substitutions executed for each load vector. Note, that in the case of multiple loads, the ψ_a matrix has more than one column. In this case, it is important to assure the linear independency of the columns.

One can extend this method to produce a pseudo-mode shape from the static solution vector and augment the computed modal space with this vector. By the nature of this augmentation the method is then called modal truncation augmentation method [1]. Note, that if the $F_a(t)$ load has multiple columns, the method described here would work with multiple residual vectors. Let us augment the modal space by the residual vector (for simplicity of discussion we consider only one)

$$\Psi_{ah+} = \begin{bmatrix} \Phi_{ah}^m & \psi_a \end{bmatrix}.$$

Here the index h^+ represents the fact that the h size has been augmented by at least one additional vector. Let us execute the modal reduction of the stiffness and the mass matrix with this mode set as

$$k_{h+h+} = \Psi_{ah+}^T K_{aa} \Psi_{ah+}$$

and

$$m_{h+h+} = \Psi_{ah+}^T M_{aa} \Psi_{ah+}.$$

Solve the modal eigenvalue problem of

$$k_{h+h+} \phi_{h+h+} = m_{h+h+} \phi_{h+h+} \Lambda_{h+h+}$$

for all the eigenvalues and eigenvectors. These will be the basis of the final mode shapes of

$$\Phi_{ah+} = \Psi_{ah+} \phi_{h+h+}.$$

The last vector in Φ_{ah+} corresponds to the residual vector shape and it is now mass and stiffness orthogonalized. Hence, executing the modal reduction with this modal space will result in improved modal solutions. See [2] for some practical results using this technique.

The effect of the application of residual vectors is apparent in the following example. The applied load was to simulate a multiple G-force impact with load pattern shown in Figure 13.1. The duration of the impact load was 0.005 seconds.

The importance of correctly identifying such peak response locations is the most challenging component of modal analysis techniques. Missing such peak

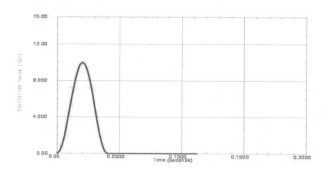

FIGURE 13.1 Time dependent load

response components could result in catastrophic failures in industrial behavior of structures.

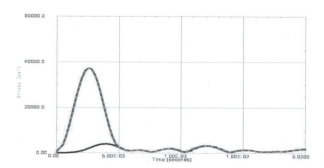

FIGURE 13.2 The effect of residual vector

The solid line shown in Figure 13.2 represents the results without using the residual vector. The dotted line represents the response with the inclusion of the residual vector. It is clearly following the peak of the load in the time interval.

13.4 The method of mode acceleration

There is yet another class of methods to improve the accuracy of modal solutions called the mode acceleration method [1]. This method is applied *a posteriori* to improve the modal solution.

Let us reorganize the dynamic response equation as

$$K_{aa}v(t) = F(t) - M_{aa}\ddot{v}(t) - B_{aa}\dot{v}(t).$$

Furthermore, solve for the displacement response

$$v(t) = K_{aa}^{-1}F(t) - K_{aa}^{-1}M_{aa}\ddot{v}(t) - K_{aa}^{-1}B_{aa}\dot{v}(t).$$

Let us assume that the modal acceleration and velocity of the system are well represented by the computed modal space:

$$\dot{v}(t) = \Phi_{ah}^{m}\dot{w}(t)$$

and

$$\ddot{v}(t) = \Phi_{ah}^{m}\ddot{w}(t).$$

Approximate the inverse of the stiffness matrix with the computed modal space as

$$K_{aa}^{-1} \approx \Phi_{ah}^{m}\Lambda^{-1,m}\Phi_{ah}^{m,T}.$$

Using the mass orthogonality property of

$$\Phi_{ah}^{m,T}M_{aa}\Phi_{ah}^{m} = I$$

and substituting the approximate inverse on the last two terms of the right hand side of the solution equation yields

$$v(t) = K_{aa}^{-1}F(t) - \Phi_{ah}^{m}\Lambda^{-1,m}\ddot{w}(t) - \Phi_{ah}^{m}\Lambda^{-1,m}B_{hh}\dot{w}(t),$$

where

$$B_{hh} = \Phi_{ah}^{m,T}B_{aa}\Phi_{ah}^{m},$$

is the modal damping matrix.

It is possible to prove that this method produces mathematically identical solution to that of the method of residual flexibility. The only difference is in the order of operations and computational complexity. The advantage of one method over the other one is problem dependent and it will not be further explored here.

13.5   Coupled modal solution application

A practical application problem for the modal solution technique is related to the structural acoustic problem presented in Chapter 6. In this application one executes two independent real symmetric eigensolutions. The first one assumes the structure is in a vacuum (fluid effects are ignored):

$$K_s \Phi_{sq_s} = M_s \Phi_{sq_s} \Lambda_s.$$

The second problem assumes that the structure in connection with the fluid is rigid:

$$K_f \Phi_{fq_f} = M_f \Phi_{fq_f} \Lambda_f.$$

The notation uses the earlier convention of q_s, q_f being the number of eigenvectors extracted from the structural and fluid problem, respectively. Let both sets of eigenvectors be mass normalized as

$$\Phi_{sq_s}^T M_s \Phi_{sq_s} = I_{q_s q_s}$$

and

$$\Phi_{fq_f}^T M_f \Phi_{fq_f} = I_{q_f q_f}.$$

Applying the modal reduction to the coupled problem yields

$$\begin{bmatrix} I_{q_s q_s} & 0 \\ \Phi_{fq_f}^T A \Phi_{sq_s} & I_{q_f q_f} \end{bmatrix} \begin{bmatrix} \ddot{w}_{q_s} \\ \ddot{w}_{q_f} \end{bmatrix} + \begin{bmatrix} \Lambda_{q_s q_s} & -\Phi_{sq_s}^T A^T \Phi_{fq_f} \\ 0 & \Lambda_{q_f q_f} \end{bmatrix} \begin{bmatrix} w_{q_s} \\ w_{q_f} \end{bmatrix} = \begin{bmatrix} \Phi_{sq_s}^T F_s \\ \Phi_{fq_f}^T F_f \end{bmatrix}.$$

Here $\Lambda_{q_s q_s}$ and $\Lambda_{q_f q_f}$ are the generalized stiffnesses. This is the coupled modal equilibrium equation that may be subject to the time integration technique shown in Chapter 14. The final results are recovered from the modal coordinates as

$$u_s = \Phi_{sq_s} w_{q_s}$$

and

$$u_f = \Phi_{fq_f} w_{q_f}.$$

The formulation shown here utilizes the fact that the constituent matrices are symmetric, although they are combined in a form that leads to unsymmetric coupled matrices.

13.6   Modal contributions and energies

Upon completion of a modal solution the engineer is faced with the problem of evaluating the contribution of the various mode shapes to the result. Identifying these modal contributions is instrumental in improving the design, for example by pin-pointing the locations where vibration absorber or other kinds of damping components may be needed in the structure [3].

The modal contributions are computed from the modal space used to generate the modal solution. Let us partition the eigenvector matrix as follows:

$$\Phi_{ah} = \begin{bmatrix} \phi_{11} & \phi_{12} & \cdots & \phi_{1i} & \cdots & \phi_{1h} \\ \phi_{21} & \phi_{22} & \cdots & \phi_{2i} & \cdots & \phi_{2h} \\ \cdots & \cdots & \cdots & \cdots & \cdots & \cdots \\ \phi_{i1} & \phi_{i2} & \cdots & \phi_{ii} & \cdots & \phi_{ih} \\ \cdots & \cdots & \cdots & \cdots & \cdots & \cdots \\ \phi_{a1} & \phi_{a2} & \cdots & \phi_{ai} & \cdots & \phi_{ah} \end{bmatrix}.$$

The ith column of this matrix corresponds to the ith eigenvector of the modal space and the components of the eigenvectors are the terms of that column. Using the fact from 13.1 that the physical solutions are recovered from the modal solutions by

$$v_a(t) = \Phi_{ah} w_h(t),$$

the total modal contribution of all modes to the kth degree of freedom at time t is computed by

$$mc_k(t) = \begin{bmatrix} \phi_{k1} & \phi_{k2} & \cdots & \phi_{ki} & \cdots & \phi_{kh} \end{bmatrix} \begin{bmatrix} w_{h,1}(t) \\ w_{h,2}(t) \\ \cdots \\ w_{h,i}(t) \\ \cdots \\ w_{h,h}(t) \end{bmatrix}.$$

Here the term $w_{h,i}(t)$ is the i-th component of the modal solution vector $w_h(t)$. The individual terms in the result of this multiplication

$$mc_{k,i}(t) = \phi_{ki}w_{h,i}(t)$$

represent the modal contribution of the ith mode to the kth physical displacement response at time t.

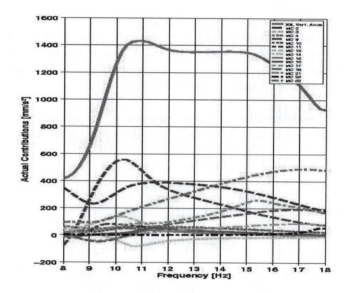

FIGURE 13.3 Modal contributions

In order to be able to compare the contribution of the individual modes, they are normalized by the total contribution of all modes as

$$mc_{k,i}^n(t) = mc_{k,i}(t)/mc_k(t).$$

Since these computations are executed at distinct time steps of the solution, their normalization base is different and additional normalization may be necessary to compute the actual contributions comparable between time instances.

These computations may also be executed at distinct excitation frequencies when the modal solution is executed in the frequency domain. This case is shown in Figure 13.3 from an auto industry example of modal contributions. The additional normalization across frequencies was executed to compute comparable actual modal contributions. The dominance of certain modes' contribution at a certain frequency enables the engineer to optimize the structure.

Another similar analysis capability with diagnostic value is the computation of modal energies. Assuming that the modal solution is result of a harmonic excitation, the modal displacement at a time instance may be expressed as

$$w_h(t) = w_h^{Re}\cos(\omega t) - w_h^{Im}\sin(\omega t)).$$

The value of analyzing the behavior of the structure via modal energies and subject to harmonic excitation is in its indicative nature for more generic type excitations of the life cycle of the product. The modal strain energy of the ith mode is computed as

$$MSE_i(t) = \frac{1}{2}[(w_{h,i}^{Re,T}\cos(\omega t) - w_{h,i}^{Im,T}\sin(\omega t))k_{ii}(w_{h,i}^{Re}\cos(\omega t) - w_{h,i}^{Im}\sin(\omega t))],$$

where the k_{ii} term is the $i$the modal stiffness and $w_{h,i}$ is the ith modal displacement component. Arithmetic manipulations yield the instructive form of

$$MSE_i(t) = \frac{1}{4}k_{ii}((w_{h,i}^{Re})^2 + (w_{h,i}^{Im})^2)+$$

$$\frac{1}{4}k_{ii}((w_{h,i}^{Re})^2 - (w_{h,i}^{Im})^2)\cos(2\omega t) - \frac{1}{2}k_{ii}w_{h,i}^{Re}w_{h,i}^{Im}\sin(2\omega t).$$

The first term is the constant and the second two are the oscillating portions of the modal strain energy. The computed energy may be presented as

$$MSEe^{i2\omega t} = MSE_{const} + MSE_{osc}e^{i2\omega t}.$$

Modal kinetic energy may be computed via an identical process, but replacing k_{ii} by the ith modal mass m_{ii} and replacing the modal solution terms $w_{h,i}$ with the modal velocities $\dot{w}_{h,i}$ resulting in

$$MKE_i(t) = \frac{1}{4}m_{ii}((\dot{w}_{h,i}^{Re})^2 + (\dot{w}_{h,i}^{Im})^2)+$$

$$\frac{1}{4}m_{ii}((\dot{w}_{h,i}^{Re})^2 - (\dot{w}_{h,i}^{Im})^2)\cos(2\omega t) - \frac{1}{2}m_{ii}\dot{w}_{h,i}^{Re}\dot{w}_{h,i}^{Im}\sin(2\omega t).$$

The structure of the modal kinetic energy is the same

$$MKEe^{i2\omega t} = MKE_{const} + MKE_{osc}e^{i2\omega t}.$$

Figure 13.4 enables the engineer to observe various scenarios. In the figure the dotted line represents the total modal kinetic energy. On the left, modes

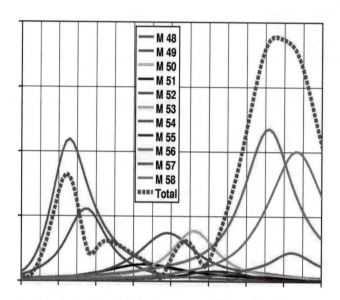

FIGURE 13.4 Modal kinetic energy distribution

48 and 49 are out of phase and result in a lower total energy than the higher mode 48. On the contrary, on the right hand side, modes 57 and 58 are in phase, hence the total modal energy is higher in their region.

References

[1] Dickens, J. M. Nakagawa, J. M. and Wittbrodt, M. J.; A critique of mode acceleration and modal truncation methods for modal response analysis, Computers and Structures, Vol. 62, No.6, pp. 985-998, 1997

[2] Rose, T.; Using residual vectors in MSC/NASTRAN dynamic analysis to improve accuracy, Proceedings of NASTRAN World User's Conference, The MacNeal-Schwendler Corporation, Los Angeles, 1991

[3] Wamsler, M.; The role of actual modal contributions in the optimization of structures, Engineering with Computers, Springer, August, 2008

14

Transient Response Analysis

We now address the problem of calculating the displacement response of the mechanical system in the time domain, called transient response analysis. The transient response analysis is the solution of

$$M\ddot{v} + B\dot{v} + Kv(t) = F(t).$$

Note, that the subscript of the matrices indicating their reduction state is omitted as the techniques discussed here are applicable to either a-size or h-size matrices. In the first case it is called direct transient, in the second case modal transient solution. Naturally, the computational costs are significantly less in the latter case.

This equation is solved numerically at discrete time intervals using various numerical differentiation schemes. The two major classes of computations are explicit and implicit schemes.

14.1 The central difference method

The method is based on the equidistant 3-point central difference formula for the first and second order numerical derivatives as

$$\dot{v}(t) = \frac{1}{2\Delta t}(v(t + \Delta t) - v(t - \Delta t)),$$

and

$$\ddot{v}(t) = \frac{1}{\Delta t^2}(v(t + \Delta t) - 2v(t) + v(t - \Delta t).$$

In the central difference method the equilibrium of the system is considered at time t and the displacements are calculated at time $t + \Delta t$, where the Δt is the equidistant time step, hence this is an explicit time integration scheme. Substituting into the equilibrium equation at t we get

$$M\frac{1}{\Delta t^2}[v(t+\Delta t)-2v(t)+v(t-\Delta t)]+B\frac{1}{2\Delta t}[v(t+\Delta t)-v(t-\Delta t)]+Kv(t) = F(t).$$

Reordering yields

$$[\frac{1}{\Delta t^2}M + \frac{1}{2\Delta t}B]v(t + \Delta t) =$$

$$F(t) + [\frac{2}{\Delta t^2}M - K]v(t) - [\frac{1}{\Delta t^2}M - \frac{1}{2\Delta t}B]v(t - \Delta t).$$

Note, that the right-hand side contains only components at time t and $t - \Delta t$. Assigning

$$C_1 = \frac{1}{\Delta t^2}M + \frac{1}{2\Delta t}B,$$

$$C_2 = F(t),$$

$$C_3 = \frac{2}{\Delta t^2}M - K,$$

and

$$C_4 = -\frac{1}{\Delta t^2}M + \frac{1}{2\Delta t}B,$$

the problem becomes

$$C_1 v(t + \Delta t) = C_2 + C_3 v(t) + C_4 v(t - \Delta t).$$

In the modal transient solution the matrices are diagonal and the problem is easy to solve. In the case of direct transient response a linear system solution is required at each time step.

Assuming that the M, B and K matrices are constant (do not change with time) and the time step Δt is also constant, the C_1 matrix needs to be factored only once. Otherwise the C_1 matrix needs to be factored at each time step, rendering the method very expensive.

14.2 The Newmark method

The time integration method preferred for large scale linear analyses is based on the classical Newmark method. The foundation of the Newmark-β method [3] is the following approximation

$$v(t + \Delta t) = \beta v(t) + (1 - 2\beta)v(t + \Delta t) + \beta v(t + 2\Delta t).$$

As the equation contains terms at time $t + \Delta t$ on both sides, it is an implicit method. The central difference forms from the previous section computed at time $t + \Delta t$ are

$$\dot{v}(t + \Delta t) = \frac{1}{2\Delta t}(v(t + 2\Delta t) - v(t)),$$

and

$$\ddot{v}(t + \Delta t) = \frac{1}{\Delta t^2}(v(t + 2\Delta t) - 2v(t + \Delta t) + v(t)).$$

Considering the equilibrium equation at time $t + \Delta t$

$$M\ddot{v}(t + \Delta t) + B\dot{v}(t + \Delta t) + Kv(t + \Delta t) = F(t + \Delta t).$$

By substituting we obtain

$$\frac{M}{\Delta t^2}[v(t + 2\Delta t) - 2v(t + \Delta t) + v(t)] + \frac{B}{2\Delta t}[v(t + 2\Delta t) - v(t)]+$$
$$K[\beta v(t + 2\Delta t) + (1 - 2\beta)v(t + \Delta t) + \beta v(t)] = F(t + \Delta t).$$

The stability of this formulation will be discussed in more detail in the next section. The choice of $\beta = \frac{1}{3}$ originally recommended by Newmark renders the form unconditionally stable. Using this value and reorganizing one obtains

$$[\frac{1}{\Delta t^2}M + \frac{1}{2\Delta t}B + \frac{1}{3}K]v(t + 2\Delta t) = F(t + \Delta t)+$$
$$[\frac{2}{\Delta t^2}M - \frac{1}{3}K]v(t + \Delta t) + [\frac{-1}{\Delta t^2}M + \frac{1}{2\Delta t}B - \frac{1}{3}K]v(t).$$

Some practical implementations average the load vector over 3 time steps also:

$$F(t + \Delta t) = \frac{1}{3}(F(t + 2\Delta t) + F(t + \Delta t) + F(t)).$$

Introducing intermediate matrices we have the following integration scheme

$$C_1 v(t + 2\Delta t) = C_2 + C_3 v(t + \Delta t) + C_4 v(t),$$

where the coefficients are:

$$C_1 = \frac{M}{\Delta t^2} + \frac{B}{2\Delta t} + \frac{K}{3},$$

$$C_2 = \frac{1}{3}(F(t + 2\Delta t) + F(t + \Delta t) + F(t)),$$

$$C_3 = \frac{2M}{\Delta t^2} - \frac{K}{3},$$

and

$$C_4 = -\frac{M}{\Delta t^2} + \frac{B}{2\Delta t} - \frac{K}{3}.$$

An iteration process is executed by factoring the C_1 matrix and solving against the current right-hand side. In essence each time step is a static solution, however, the right-hand side is updated at each time step.

If we assume again that the M, B and K matrices are constant and do not change with time and the time step Δt is constant, then the C_1 matrix needs to be factored only once. If either one of these conditions is untrue, the C_1 matrix needs to be factored at each time step, a daunting task indeed.

14.3   Starting conditions and time step changes

We now need to discuss the starting conditions of these algorithms. If we start at time $t = 0$ and the first step is to evaluate the equilibrium at time $t = \Delta t$, we need to have $v(0), F(0), v(-\Delta t), F(-\Delta t)$ specified.

Let us assume that initial displacement and velocity values are specified by the engineer. The starting components are computed from the assumption that at $t < 0$ the acceleration of the system is zero. Then, the negative time components are computed as

$$v(-\Delta t) = v(0) - \dot{v}(0)\Delta t,$$

and

$$F(-\Delta t) = Kv(-\Delta t) + B\dot{v}(0).$$

The starting load is also computed to assure that the acceleration at $t = 0$ is also zero:

$$F(0) = Kv(0) + B\dot{v}(0).$$

It is sometimes necessary to change the time step in the middle of the process. Naturally, in this case the C_1 matrix needs to be re-factored. In addition, the starting components of the new time integration process need to be computed.

Let us consider a case of stopping an integration process at time $t = T$ and change the time step. The initial conditions for the new process are clearly

$$v(0) = v(T), \quad F(0) = F(T)$$

and

$$\dot{v}(0) = \dot{v}(T), \quad \ddot{v}(0) = \ddot{v}(T).$$

The starting components for the new integration process starting at time $t = T$ with time step Δt_2 are computed as

$$v(-\Delta t_2) = v(T) - \dot{v}(T)\Delta t_2 + \frac{1}{2}\Delta t_2^2 \ddot{v}(T)$$

and

$$F(-\Delta t_2) = Kv(-\Delta t_2) + B[\dot{v}(T) - \Delta t_2 \ddot{v}(T)] + M\ddot{v}(T).$$

14.4 Stability of time integration techniques

An issue of consequence is the stability of these numerical methods. A stable numerical solution method has the characteristic of producing small changes in the subsequent approximation steps if the initial conditions are subjected to small changes.

A two-step numerical solution scheme of

$$w_{i+1} = f_i + bw_i + cw_{i-1}$$

has its characteristic equation defined by

$$P(\lambda) = \lambda^2 - b\lambda - c = 0.$$

If the roots of the characteristic equations satisfy

$$|\lambda_i| \leq 1$$

then the method is called stable.

For the stability analysis of the two methods discussed above, we consider the modal form of the time domain equilibrium equation. Without restricting the generality of the following discussion, we assume that the full modal space was computed during the modal formulation. This results in a completely decoupled set of scalar differential equation of form:

$$\ddot{w}(t) + 2\xi_i \omega_i \dot{w}(t) + \omega_i^2 w(t) = f_i(t), \quad i = 1,..n.$$

Here

$$\Phi_i^T B \Phi_i = 2\xi_i \omega_i$$

implies proportional modal damping and

$$f_i(t) = \Phi_i F(t).$$

The orthogonality conditions produce

$$\Phi_i^T K \Phi_i = \omega_i^2$$

and

$$\Phi_i^T M \Phi_i = 1.$$

The stability analysis of the i-th decoupled equation enables drawing conclusions for any particular method.

The central difference method in that context is formulated as

$$w(t + \Delta t)[1 + \xi_i\omega_i\Delta t] = f_i(t)\Delta t^2 + w(t)[2 - \omega_i^2\Delta t^2] - w(t - \Delta t)[1 - \xi_i\omega_i\Delta t].$$

The characteristic equation for this case is

$$\lambda^2(1 + t) + (r - 2)\lambda + (1 - t) = 0,$$

where

$$t = \xi_i\omega_i\Delta t$$

and

$$r = \omega_i^2\Delta t^2.$$

Reformulating into a form with unit leading coefficient

$$\lambda^2 + \frac{r - 2}{1 + t}\lambda + \frac{1 - t}{1 + t} = 0$$

yields the solution of the characteristic equation as

$$\lambda_{1,2} = \frac{2 - r}{2(1 + t)} \mp \frac{1}{2}\sqrt{(\frac{r - 2}{1 + t})^2 - 4\frac{1 - t}{1 + t}}.$$

Setting $\lambda_{1,2} = 1$ and simplifying with the $t \neq 0$ condition one obtains the limit of

$$\Delta t = \frac{2\xi_i}{\omega_i}.$$

Introducing the period of the vibration $(T_i = \frac{2\Pi}{\omega_i})$ instead of the frequency

$$\Delta t = \frac{T_i\xi_i}{\Pi}.$$

The interpretation of this is that for stability we need to impose a limit on the size of the time step (Δt). The central difference method hence is a conditionally stable method, above the critical value of the time step the method is unstable. The condition for the complete model is

$$\Delta t_{critical} = \frac{cT}{\Pi},$$

where T is the smallest free vibration period of the finite element model and c is a constant [1]. As this value is not necessarily or easily available, especially in the case of unreduced (a-set size) solution, the explicit method is the preferred method only in cases when the time step can be kept very small to assure stability. An example of such cases is the the crash analysis of automobiles.

The stability of the Newmark method is discussed next. Using the same approach as above, the Newmark equilibrium equation may be written as

$$w(t + 2\Delta t)[1 + \xi_i \omega_i \Delta t + \beta \omega_i^2 \Delta t^2] =$$

$$w(t + \Delta t)[2 - (1 - 2\beta)\omega_i^2 \Delta t^2] + w(t)[-1 + \xi_i \omega_i \Delta t - \beta \omega_i^2 \Delta t^2] + f_i(t + 2\Delta t).$$

The characteristic equation defining the stability is

$$\lambda^2(1 + t + \beta r) + \lambda(-2 + (1 - 2\beta)r) + (1 - t + \beta r) = 0.$$

Here t, r are the same as above. Rewrite the equation to unit lead coefficient

$$\lambda^2 + \lambda \frac{-2 + (1 - 2\beta)r}{1 + t + \beta r} + \frac{(1 - t + \beta r)}{1 + t + \beta r} = 0.$$

The solution of this equation is

$$\lambda_{1,2} = -\frac{-2 + (1 - 2\beta)r}{2(1 + t + \beta r)} \mp \frac{1}{2}\sqrt{(\frac{-2 + (1 - 2\beta)r}{1 + t + \beta r})^2 - 4\frac{1 - t + \beta r}{1 + t + \beta r}}.$$

Assume real solutions for simplicity of our algebra. The general complex solution produces the same final result. Setting $\lambda_{1,2} = 1$ and reordering yields

$$4 + (4\beta - 1)r = 0.$$

Assuming that $r \neq 0$ (or $\omega_i \neq 0$) the selection of $\beta \geq \frac{1}{4}$ will assure that the condition of the eigenvalues being less than unity will always be satisfied. Hence, that is a lower bound for β. The value of $\frac{1}{3}$ used in the method developed in the last paragraph is slightly higher, providing a safety cushion of stability.

In conclusion, the Newmark method with a proper selection of β is unconditionally stable, i.e., there is no limit on the time step sizes. Naturally, there are other accuracy considerations one must be aware of, for example irrationally large time steps will produce large finite difference approximation errors.

14.5   Transient response case study

The case study involves the modal transient response calculation of a complex chassis and wheel assembly of a car body of which it is impractical to show a picture here. The subject of the analysis was to establish the transient response of the model to an impact type load as shown in Figure 14.1.

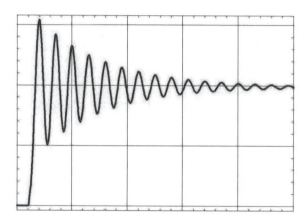

FIGURE 14.1 Transient response

The figure demonstrates the effect of the initial impact and the following stabilization of the structure indicated by the gradual decaying of the motion. The model consisted of 1.4 million nodes and over 788 thousand finite elements. The total degrees of freedom was above 8.6 million. The solution was obtained with the modal transient response approach and 180 modes up to 3,000 Hz were captured to represent the modal space.

The total transient response analysis required 2,997 minutes of elapsed time, which amounts to more than two days of computing, clearly a weekend job. Of this time 2,175 minutes were spent on the computation of the modal space (the eigenvalue solution) and the rest was the time integration as well as the engineering result computations.

The run used 315 Mwords of memory and the disk high water level was 223 GBytes. The amount of I/O executed was a huge 6.65 Terabytes. The analysis was performed on a workstation containing 4 (1.5 GHz) CPUs.

Note that this solution would have not been practical at all without the modal approach even on a higher performing workstation.

14.6 State-space formulation

In complex spectral computations in Chapter 10 it was proven practical to transfer the quadratic problem into a linear problem of twice the size. The concept applied to the transient response problem leads to the state-space formulation. This is especially advantageous when viscous damping is applied, such as occurring when modeling the shock absorber and other vibration control devices of car bodies.

We consider the transient response problem

$$M\ddot{v}(t) + B\dot{v}(t) + Kv(t) = f(t),$$

and rewrite it as a 2 by 2 block linear problem:

$$\begin{bmatrix} M & 0 \\ 0 & I \end{bmatrix} \begin{bmatrix} \ddot{v}(t) \\ \dot{v}(t) \end{bmatrix} + \begin{bmatrix} B & K \\ -I & 0 \end{bmatrix} \begin{bmatrix} \dot{v}(t) \\ v(t) \end{bmatrix} = \begin{bmatrix} f(t) \\ 0 \end{bmatrix}.$$

Inverting and reordering brings the format of

$$\begin{bmatrix} \ddot{v}(t) \\ \dot{v}(t) \end{bmatrix} = \begin{bmatrix} -M^{-1}B & -M^{-1}K \\ I & 0 \end{bmatrix} \begin{bmatrix} \dot{v}(t) \\ v(t) \end{bmatrix} = \begin{bmatrix} M^{-1}f(t) \\ 0 \end{bmatrix}.$$

Introducing the so-called state vector

$$x = \begin{bmatrix} \dot{v}(t) \\ v(t) \end{bmatrix}$$

and its derivative

$$\dot{x} = \begin{bmatrix} \ddot{v}(t) \\ \dot{v}(t) \end{bmatrix}$$

results in the state-space formulation of

$$\dot{x} = Ax + u.$$

The state transition matrix is the consequence of above:

$$A = \begin{bmatrix} -M^{-1}B & -M^{-1}K \\ I & 0 \end{bmatrix},$$

and the state input vector is

$$u = \begin{bmatrix} M^{-1}f(t) \\ 0 \end{bmatrix}.$$

This formulation appears to be simpler to solve than the time integration techniques presented earlier in this chapter. This, however, requires the existence of the inverse of the M matrix and the solution of the indefinite system may also pose computational difficulties.

The modal transient problem may also be brought to a state-space form. Let the modal reduction of the undamped problem represented by

$$\omega_i = \phi_i^T K \phi_i,$$

$$2\xi_i \omega_i = \phi_i^T B \phi_i,$$

and

$$\phi_i^T M \phi_i = 1,$$

for $i = 1, 2, \ldots, h$. Then the modal state transition matrix is formed as

$$a = \begin{bmatrix} -2\xi_1\omega_1 & 0 & 0 & -\omega_1^2 & 0 & 0 \\ 0 & -2\xi_i\omega_i & 0 & 0 & -\omega_i^2 & 0 \\ 0 & 0 & -2\xi_h\omega_h & 0 & 0 & -\omega_h^2 \\ & I_{hh} & & & 0_{hh} & \end{bmatrix}.$$

The modal input becomes

$$u_h = \begin{bmatrix} \Phi_h^T M^{-1} f(t) \\ 0 \end{bmatrix}.$$

Introducing the modal displacement

$$v = \Phi_h w$$

and modal velocity of

$$\dot{v} = \Phi_h \dot{w},$$

the modal state variable becomes

$$x_h = \begin{bmatrix} \dot{w}(t) \\ w(t) \end{bmatrix},$$

with derivative

$$\dot{x}_h = \begin{bmatrix} \ddot{w}(t) \\ \dot{w}(t) \end{bmatrix}.$$

The modal state-space form is

$$\dot{x}_h = ax_h + u_h.$$

In practical applications the input load may only affect a few degrees of freedom of the system and conversely, the results may also be only needed at a few locations. These locations are not necessarily the same, leading to the so-called transfer mobility problem: What is the effect of a load applied to one place of the structure to another location?

For example, the effect of the wheel excitation to the driver seat vertical acceleration is a standard evaluation procedure in the NVH (Noise, Vibration and Harshness) analysis in the automobile industry. By introducing a B input coupling matrix and the C output selection matrix, the formulation may be refined as a pair of equations

$$\dot{x} = Ax + Bu,$$

$$y = Cx.$$

Here y contains the selected output. For example, if the engineer wants to measure the response only at a single output location, the C matrix will have a single row and as many columns as the state vector size (twice the free degrees of freedom in the system). The terms of the matrix will be all zero, except for the location corresponding to the output degree of freedom.

Conversely, if the input load affects only one degree of freedom of the model, the B matrix will have a single column with all zeroes, but for the location of the loaded degree of freedom which will be one. Multiple loaded or sensed degrees of freedom will result in multiple rows and columns of the C and B matrices, respectively.

Similar form for the modal case is also possible,

$$\dot{x}_h = ax_h + bu_h$$

and

$$y = cx_h,$$

but special considerations are necessary in populating b and c accommodating the translation between the physical and modal degrees of freedom. A prominent application of the state-space formulation is in analyzing structures cooperating with control systems or active damping components [2].

References

[1] Bathe, K.-J. and Wilson, E. L.; Stability and accuracy analysis of direct integration methods, Int. Journal of Earthquake Engineering and Structural Dynamics, Vol. 1, pp. 283-191, 1973

[2] Gawronski, W. K.; Balanced control of flexible structures, Springer, New York, 1996

[3] Newmark, N. M.; A method of computation for structural dynamics, Proceedings of ASME Conference, ASME, 1959

15

Frequency Domain Analysis

Now we address the problem of calculating the displacement response of the mechanical system in the frequency domain. These calculations consider a load exerted on the structure with a frequency-dependent component. Since the computation is in the frequency domain, it is called frequency response analysis.

The discussions below focus on the undamped symmetric case, although the methodologies directly carries over to the more general cases of the occurrence of unsymmetric matrices [1] and the damped (3 matrix) problem [2].

15.1 Direct and modal frequency response analysis

The frequency response solution methodology requires the calculation of the response of a mechanical model to external excitation in a sometimes very wide frequency range. That is expensive and the computational resource requirements are significant.

The direct frequency response of an undamped structure at an excitation frequency ω_j is described by

$$(K_{aa} - \omega_j^2 M_{aa})u_a(\omega_j) = F_a(\omega_j).$$

Here u is the response vector and F is the external load. The direct method provides a computationally exact solution for the engineering problem at hand.

The modal frequency response of an undamped structure is described by

$$(K_{hh} - \omega_j^2 M_{hh})u_h(\omega_j) = F_h(\omega_j).$$

The modal frequency response solution provides a computationally exact solution to the engineering problem only when the full modal space is used.

Note, that the loads may be frequency-dependent and the equation is evaluated at many given frequency locations $\omega_j, j = 1, .., m$. Therefore, the solution is also frequency-dependent.

The solution techniques of this chapter are applicable to both of these cases, therefore, the subscripts will be deliberately ignored in the remainder of the chapter.

Let us introduce $\mu = \omega_j^2$ for simplicity. Let us assume for now also that the matrices are symmetric; this is not a restriction of generality, it just clarifies the discussion. Then the direct response at the j-th frequency is given as the solution of

$$(K - \mu M)u(\mu) = F(\mu).$$

For numerical stability the problem is sometimes shifted, or a form of spectral transformation is executed as follows. By defining a shift $\sigma \neq \mu$ in the neighborhood of μ and pre-multiplying this equation we get

$$(K - \sigma M)^{-1}(K - \mu M)u(\mu) = (K - \sigma M)^{-1}F(\mu).$$

In short, this is called the shifted dynamic equation of the response problem

$$D(\mu)u(\mu) = q(\mu),$$

where the shifted dynamic matrix is

$$D(\mu) = (K - \sigma M)^{-1}(K - \mu M),$$

and the modified load is

$$q(\mu) = (K - \sigma M)^{-1}F(\mu).$$

The direct solution is then based on the

$$D(\mu) = LU$$

factorization and the following forward-backward substitution. These techniques were shown in Chapter 7.

15.2   Reduced-order frequency response analysis

The reduced-order modeling method enables the approximate solution of the frequency response problem with much reduced resource requirements. The reduced-order response technique is an alternative spectral reduction technique. In this case, the spectrum of the dynamic matrix is approximated by a Krylov subspace as opposed to approximation by actual eigenvectors.

The shifted dynamic matrix may be written as

$$D(\mu) = (K - \sigma M)^{-1}(K - (\sigma + \mu - \sigma)M) = I + (\sigma - \mu)S,$$

where

$$S = (K - \sigma M)^{-1}M.$$

A k-th order Krylov subspace spanned by this matrix S is defined by

$$\kappa_k = span(q, Sq, S^2q, S^3q, ..., S^{k-1}q),$$

where q is a starting vector.

To generate the basis vectors for the Krylov subspace, the Lanczos algorithm, first introduced in Chapter 9, is an excellent candidate. Starting from $v_1 = q/\beta, \beta = ||q||$ the Lanczos method produces a set of vectors

$$V_k = \begin{bmatrix} v_1 & v_2 & v_3 & ... & v_k, \end{bmatrix}$$

that span the k-th order Krylov subspace of S:

$$\kappa_k = span(v_1, v_2, v_3, ..., v_k).$$

The Lanczos recurrence is described by

$$SV_k - V_kT_k = \beta_k v_{k+1}e_k^T,$$

where the T_k matrix is tridiagonal containing the orthogonalization and normalization parameters of the process. The V_k vectors are orthonormal: $V_k^T V_k = I$. The right-hand side term represents the truncation error in case the subspace size k is less then the matrix size n.

Note, that if the matrices are not symmetric the bi-orthogonal Lanczos method from Section 10.2 is used resulting in two (bi-orthogonal) sets of Lanczos vectors. For the sake of the following derivation, let us ignore the right-hand side term temporarily. Reordering and pre-multiplying yields

$$(\sigma - \mu)SV_k = (\sigma - \mu)V_kT_k.$$

From the reordered form of the shifted dynamic matrix it follows that

$$(\sigma - \mu)S = D(\mu) - I.$$

Substituting the latter into the prior equation produces

$$(D(\mu) - I)V_k = (\sigma - \mu)V_k T_k.$$

Another reordering yields

$$D(\mu)V_k = V_k(I + (\sigma - \mu)T_k).$$

Introducing

$$u(\mu) = V_k \bar{u}(\mu),$$

and substituting into the dynamic equation results in

$$V_k(I + (\sigma - \mu)T_k)\bar{u}(\mu) = q(\mu).$$

Finally, pre-multiplying by V_k^T produces

$$(I + (\sigma - \mu)T_k)\bar{u}(\mu) = V_k^T q(\mu).$$

The last equation is a reduced (k-th) order problem (hence the name of the method) as

$$\overline{D}_k(\mu)\bar{u}(\mu) = \bar{q}(\mu).$$

Here

$$\overline{D}_k(\mu) = (I + (\sigma - \mu)T_k),$$
$$\bar{u}(\mu) = V_k^T u(\mu),$$

and

$$\bar{q} = V_k^T q.$$

This also may be considered the projection of our original problem onto the Krylov subspace.

The reduced-order problem (which is usually not formed explicitly) may be solved very conveniently. Solving for $u(\mu)$ produces the final response solution

$$u(\mu) = V_k(I + (\sigma - \mu)T_k)^{-1}V_k^T q(\mu).$$

In the case of constant right-hand side, $q(\mu) = q(\sigma)$, a further simplification is possible. Since $q = \beta v_1$ and $V_k^T V_k = I$, the right-hand side reduces to

$$V_k^T q = \beta e_1,$$

where e_1 is the first unit vector of the k-dimensional subspace. In this case, the final response solution is simply

$$u(\mu) = \beta V_k(I + (\sigma - \mu)T_k)^{-1}e_1.$$

Two specific aspects of this equation are noteworthy. One is that the matrix to be inverted here is tridiagonal. That, of course, is a computational advantage, ignoring the cost of producing the tridiagonal matrix T_k for the moment. Secondly, appropriately chosen σ enables finding approximate solutions at several μ locations without recomputing T_k, V_k.

Finally, since the order of the Krylov subspace is less than that of the original dynamic equation ($k < n$), the reduced-order, approximate solution acceptance issue needs to be addressed.

15.3   Accuracy of reduced-order solution

Revisiting the Lanczos recurrence equation above, including the truncation term and adding a term accounting for round-off error in finite arithmetics we obtain

$$SV_k - V_kT_k = \beta_k v_{k+1}e_k^T + E_k,$$

where $||E_k|| = \epsilon_{machine}||S||$ and $\epsilon_{machine}$ is representative of the floating point arithmetic accuracy. The residual of the original, full-size response problem is

$$r(\mu) = D(\mu)u(\mu) - q(\mu).$$

Combining the last two equations and repeating some algebraic steps, we produce a form useful for convergence estimate

$$r(\mu) = (\sigma - \mu)(\beta_k v_{k+1}e_k^T + E_k)\bar{q}.$$

In practice the relative error of

$$e(\mu) = \frac{||r(\mu)||}{||F(\mu)||}$$

is used, where

$$||r(\mu)|| = |\sigma - \mu|max(\ \beta_k|e_k^T\bar{q}|\ ;\ ||E_k||\ ||\bar{q}||\ ).$$

Finally, a response at a certain frequency is accepted if

$$e(\mu) \leq \epsilon_{acceptance},$$

where $\epsilon_{acceptance}$ is based on various engineering criteria.

It should be noted that the technique is advantageous mainly in models dominated by three-dimensional (solid) elements. These models have more widely spaced natural frequencies. Such models arise in the analysis of automobile components such as brakes and engine blocks.

The accuracy of the approximated responses deteriorates quickly if there are natural frequencies between the shift point and the response points. These issues unfortunately limit the performance of this method in response analysis of shell structures. The latter class include car body and airplane fuselage or wing models.

15.4   Frequency response case study

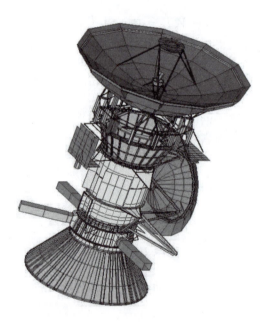

FIGURE 15.1 Satellite model

We will consider an example of a satellite similar to the one shown in Figure 15.1 as illustration. Such objects undergo a wide frequency range of excitation during their launching and operational life-cycle, therefore the frequency response analysis is of utmost importance.

A satellite model consisted of approximately 710,000 node points and 760,000 elements of various kind, reflecting the complexity of the model. The total degrees of freedom exceeded 4.2 million and the model was analyzed at 800 excitation frequencies.

The CPU time of the modal frequency response analysis was approximately 200 minutes of elapsed time on a workstation with 4 (1.5 GHz) processors. The amount of I/O operations was just over one Terabyte and almost 100 Gigabytes of disk footprint was required to complete the analysis.

15.5 Enforced motion application

It is common in engineering practice to have a non-stationary excitation on the structure called an enforced motion, as it may be displacement, velocity or even acceleration. The practical values of such approach are immense [4].

The excitation is restricted to a partition of the problem's degrees of freedom denoted by (s). Since the mechanism of applying an enforced displacement is similar to single point constraints, the notation and partitioning resembles to that process described in Section 4.5.

$$\begin{bmatrix} M_{ff} & M_{fs} \\ M_{sf} & M_{ss} \end{bmatrix} \begin{bmatrix} \ddot{v}_f \\ \ddot{v}_s \end{bmatrix} + \begin{bmatrix} B_{ff} & B_{fs} \\ B_{sf} & B_{ss} \end{bmatrix} \begin{bmatrix} \dot{v}_f \\ \dot{v}_s \end{bmatrix} + \begin{bmatrix} K_{ff} & K_{fs} \\ K_{sf} & K_{ss} \end{bmatrix} \begin{bmatrix} v_f \\ v_s \end{bmatrix} = \begin{bmatrix} P_f \\ 0_s \end{bmatrix}.$$

The commonly used approach is based on the first equation as

$$M_{ff}\ddot{v}_f + B_{ff}\dot{v}_f + K_{ff}v_f = P_f - (M_{fs}\ddot{v}_s + B_{fs}\dot{v}_s + K_{fs}v_s).$$

Note that the right hand side contains the active load and the enforced motion terms and as such, properly computable [3]. We assume a frequency dependent enforced motion of the structure in the form of

$$v_s(t) = u_s(\omega)e^{i\omega t}$$

resulting in a harmonic response

$$v_f(t) = u_f(\omega)e^{i\omega t}.$$

It follows that

$$\dot{v}_f = i\omega u_f, \quad \ddot{v}_f = -\omega^2 u_f,$$

and

$$\dot{v}_s = i\omega u_s, \quad \ddot{v}_s = -\omega^2 u_s.$$

Substituting into the equilibrium equation produces the governing equation in the frequency domain

$$(-\omega^2 M_{ff} + i\omega B_{ff} + K_{ff})u_f = P_f - (-\omega^2 M_{fs} + i\omega B_{fs} + K_{fs})u_s.$$

Introducing

$$Z_{ff} = -\omega^2 M_{ff} + i\omega B_{ff} + K_{ff},$$

$$Z_{fs} = -\omega^2 M_{fs} + i\omega B_{fs} + K_{fs},$$

$$Z_{sf} = -\omega^2 M_{sf} + i\omega B_{sf} + K_{sf},$$

and

$$Z_{ss} = -\omega^2 M_{ss} + i\omega B_{ss} + K_{ss}.$$

the complete frequency domain equation of the enforced motion problem may be written as

$$\begin{bmatrix} Z_{ff} & Z_{fs} \\ Z_{sf} & Z_{ss} \end{bmatrix} \begin{bmatrix} u_f \\ u_s \end{bmatrix} = \begin{bmatrix} P_f \\ 0 \end{bmatrix}.$$

Analyzing enforced motion in the modal space requires that the static shapes associated with the unit motion of each enforced motion point be appended to the flexible mode shapes when doing modal reduction. These static shapes are computed as

$$\Phi_{fs} = -K_{ff}^{-1} K_{fs}.$$

The flexible mode shapes are obtained from the eigenvalue solution as

$$K_{ff}\Phi_{fh} = M_{ff}\Phi_{fh}\Lambda_h.$$

Introducing

$$u_d = \begin{bmatrix} u_f \\ u_s \end{bmatrix},$$

and

$$\Phi_{dx} = \begin{bmatrix} \Phi_{fh} & \Phi_{fs} \\ 0 & I_{ss} \end{bmatrix},$$

the modal substitution is executed as

$$u_d = \Phi_{dx} w_x,$$

where

$$w_x = \begin{bmatrix} w_h \\ u_s \end{bmatrix}.$$

Here w_h is the modal displacement sought. Pre-multiplying the complete equation by Φ_{dx}^T results in the modal form of the problem.

$$\begin{bmatrix} \Phi_{fh}^T & 0 \\ \Phi_{fs}^T & I_{ss} \end{bmatrix} \begin{bmatrix} Z_{ff} & Z_{fs} \\ Z_{sf} & Z_{ss} \end{bmatrix} \begin{bmatrix} \Phi_{fh} & \Phi_{fs} \\ 0 & I_{ss} \end{bmatrix} \begin{bmatrix} w_h \\ u_s \end{bmatrix} = \begin{bmatrix} \Phi_{fh}^T P_f \\ 0 \end{bmatrix}.$$

Executing the posted multiplications and developing the first equation results in

$$\Phi_{fh}^T Z_{ff} \Phi_{fh} w_h = \Phi_{fh}^T P_f - \Phi_{fh}^T (Z_{ff} \Phi_{fs} + Z_{fs}) u_s.$$

Introducing modal matrices

$$z_{hh} = \Phi_{fh}^T Z_{ff} \Phi_{fh},$$

and

$$z_{hs} = \Phi_{fh}^T Z_{ff} \Phi_{fs} + \Phi_{fh}^T Z_{fs},$$

as well as modal loads

$$P_h = \Phi_{fh}^T P_f,$$

the modal solution may be obtained from

$$z_{hh} w_h = P_h - z_{fs} u_s.$$

The physical solution component is recovered by the relation

$$u_f = \Phi_{fh} w_h + \Phi_{fs} u_s.$$

The reaction forces at the constraints enforcing the displacements are computed as

$$Q = -\Phi_{fs}^T P_f.$$

Finally, it is quite simple to apply forces also to the enforced motion locations, by putting P_s into the zero block of the right hand sides of the equations.

References

[1] Freund, R.W.; Passive reduced-order modeling via Krylov subspace methods, Bell Laboratories Numerical Analysis Manuscript N0. 00-3-02,

March 2000

[2] Meerbergen, K.; The solution of parameterized symmetric linear systems and the connection with model reduction, SIAM Journal of Matrix Analysis and Applications, 2003

[3] Timoshenko, S.; Vibration problems in engineering, Wiley, New York, 1974

[4] Wamsler, M., Blanck, N. and Kern, G.; On the enforced relative motion inside a structure, Proceedings of the MSC 20th User's Conference, 1993

16

Nonlinear Analysis

In all earlier chapters we assumed that the load vs. displacement relationship is linear. In some applications, however, there is a nonlinear relationship between the loads and the displacements. This chapter presents the most widely used computational concepts of such solutions but will not focus on details of these issues, due to elaborate, specialized references, such as [1].

16.1   Introduction to nonlinear analysis

Nonlinear relationship between the loads and the displacements could occur in several categories. The simplest case is called material nonlinearity, and it is captured in the stress-strain matrix D. This may be due to nonlinear elasticity, plasticity or visco-elasticity of the material. This category, as the origin of the nonlinearity indicates, is closely related to the material modeling issues and as such it is application dependent.

In nonlinear analysis the stiffness matrix is composed of two distinct components as

$$K = K_l + K_{nl},$$

where the subscripts refer to linear and nonlinear, respectively. The accompanying strains are also of two components

$$\epsilon = \epsilon_l + \epsilon_{nl}.$$

The linear components K_l, ϵ_l are the conventional stiffness and strain as derived in Chapter 3. The second component is formulated similarly, however, in terms of the nonlinear matrices.

In the case of material nonlinearity

$$K_{nl} = \Sigma_e \int \int \int B^T D_{nl} B dV,$$

the D_{nl} is a nonlinear stress-strain matrix, representing the material non-linearity. Such a case is shown for example in Figure 16.1, when the strain exceeds the yield point for the material.

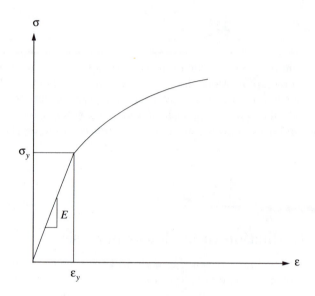

FIGURE 16.1 Nonlinear stress-strain relationship

Up to the yield point σ_y, ϵ_y, considering a bar undergoing tension or compression

$$\sigma = E\epsilon,$$

which is clearly linear. Above the yield point, however, the relationship is not linear

$$\sigma = f(E, \epsilon).$$

Finite elements formulated to accommodate this scenario are commonly called hyper-elastic elements. In the nonlinear case, due to the dependence of the stiffness matrix on the displacement, there is an imbalance between the external load and the internal forces of the model.

$$\Delta F = F - F_{int}(u).$$

F is the external force and F_{int} is the forces internal to the model computed by

$$F_{int}(u) = \Sigma_e \int \int \int B^T \sigma dV.$$

The force imbalance results in an incremental displacement of Δu as follows

$$\Delta F = K \Delta u.$$

The equilibrium of the nonlinear model is achieved when the force imbalance or the incremental displacement is zero or sufficiently small. This equilibrium is obtained by an iterative procedure that consists of steps at which the force imbalance is computed and tested. If it is not small enough, an incremental displacement related to the imbalance is computed, the displacement is adjusted and the imbalance is again evaluated. If the displacement exceeds a certain level, the stiffness matrix is also updated and the process repeated.

On a side note: in nonlinear analysis the automated singularity elimination process discussed in Chapter 5 may also be repeatedly executed. This is due to the presence of the high deformations which modify the original balance in that regard.

16.2 Geometric nonlinearity

Another case, called geometric nonlinearity, is manifested in the nonlinearity of the strain-displacement matrix B. The possible causes of this may be large deformation of some elements, or even contact between elements. This requires taking the second order displacement effects into consideration.

The nonlinear stiffness matrix in the case of geometric nonlinearity is:

$$K_{nl} = \Sigma_e \int \int \int B_{nl}^T D B_{nl} dV.$$

The non-linear strain-displacement matrix produces a nonlinear strain vector

$$\epsilon_{nl} = B_{nl} q_e.$$

The nonlinear strain vector components are of form

$$\epsilon_{nl,x} = \frac{1}{2} \left(\left(\frac{\partial q_u}{\partial x} \right)^2 + \left(\frac{\partial q_v}{\partial x} \right)^2 + \left(\frac{\partial q_w}{\partial x} \right)^2 \right)$$

and

$$\tau_{nl,xy} = (\frac{\partial q_u}{\partial x}\frac{\partial q_u}{\partial y} + \frac{\partial q_v}{\partial x}\frac{\partial q_v}{\partial y} + \frac{\partial q_w}{\partial x}\frac{\partial q_w}{\partial y}).$$

For a geometrically non-linear tetrahedral element corresponding to the linear element introduced in Section 3.4, the strain-displacement matrix producing the non-linear strain vector may be computed by introducing an 6×9 intermediate matrix of form

$$A = \begin{bmatrix} a_x^T & 0 & 0 \\ 0 & a_y^T & 0 \\ 0 & 0 & a_z^T \\ a_y^T & a_x^T & 0 \\ 0 & a_z^T & a_y^T \\ a_z^T & 0 & a_x^T \end{bmatrix},$$

with terms of

$$a_x = \begin{bmatrix} \frac{\partial q_u}{\partial x} \\ \frac{\partial q_v}{\partial x} \\ \frac{\partial q_w}{\partial x} \end{bmatrix}, \quad a_y = \begin{bmatrix} \frac{\partial q_u}{\partial y} \\ \frac{\partial q_v}{\partial y} \\ \frac{\partial q_w}{\partial y} \end{bmatrix}, \quad a_z = \begin{bmatrix} \frac{\partial q_u}{\partial z} \\ \frac{\partial q_v}{\partial z} \\ \frac{\partial q_w}{\partial z} \end{bmatrix}.$$

With the vector

$$b = \begin{bmatrix} a_x \\ a_y \\ a_z \end{bmatrix},$$

it is easy to verify that the multiplication of

$$\epsilon_{nl} = \frac{1}{2}Ab,$$

produces the geometrically nonlinear strains of above forms. Substituting the shape function derivatives as

$$b = Cq_e,$$

where the 9×12 matrix

$$C = \begin{bmatrix} \frac{\partial N_1}{\partial x} & 0 & 0\frac{\partial N_2}{\partial x} & 0 & 0\frac{\partial N_3}{\partial x} & 0 & 0\frac{\partial N_4}{\partial x} & 0 & 0 \\ 0 & \frac{\partial N_1}{\partial y} & 0 & 0\frac{\partial N_2}{\partial y} & 0 & 0\frac{\partial N_3}{\partial y} & 0 & 0\frac{\partial N_4}{\partial y} & 0 \\ 0 & 0 & \frac{\partial N_1}{\partial z} & 0 & 0\frac{\partial N_2}{\partial z} & 0 & 0\frac{\partial N_3}{\partial z} & 0 & 0\frac{\partial N_3}{\partial z} \end{bmatrix},$$

the 6×12 non-linear strain-displacement matrix becomes

$$B_{nl} = \frac{1}{2}AC,$$

satisfying

$$\epsilon_{nl} = B_{nl} q_e.$$

The computation of the non-linear element matrix now proceeds according to Section 3.4.

A possible cause of geometric nonlinearity is large rotation of some of the structural elements. Large rotations of the elements could occur in many different physical applications. For a simple example we reconsider the bar model introduced in Chapter 4 undergoing tension-compression. In this case, however, we let it also attain a large rotation around the first node point on the left as shown in Figure 16.2.

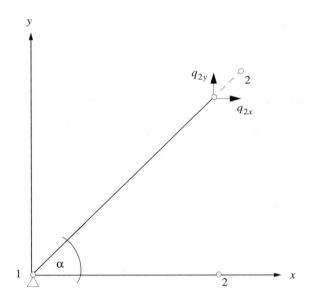

FIGURE 16.2 Rotated bar model

We will consider two degrees of freedom per node point for x and y displacements. The strain-displacement relationship is

$$\epsilon = B_{nl} q_e,$$

where

$$q_e = \begin{bmatrix} q_{1_x} \\ q_{1_y} \\ q_{2_x} \\ q_{2_y} \end{bmatrix},$$

and x is the original axial direction prior to the rotation and y is perpendicular to that. The strain-displacement matrix

$$B_{nl} = \frac{1}{l} \begin{bmatrix} -1 & 0 & 1 & 0 \\ 0 & -1 & 0 & 1 \end{bmatrix},$$

will result in the strains accommodating the large rotation as

$$\epsilon = \begin{bmatrix} l\cos(\alpha) - q_{1x} \\ l\sin(\alpha) - q_{1y} \end{bmatrix},$$

where l is the length of the bar and α is the angle or rotation.

16.3 Newton-Raphson methods

The iterative process applied in the industry is based on the well-known Newton-Raphson iteration method of finding the zero of a nonlinear equation. In that method the solution to the nonlinear equation $g(x) = 0$ is obtained by approximating the equation with its first order Taylor polynomial around a point x^i in the iteration process, where $i = 1, 2, \ldots$ until convergence is achieved.

$$g(x) = g(x^i) + g'(x^i)(x - x^i)$$

where

$$g'(x^i) = \frac{dg(x)}{dx}\Big|_{x=x^i}.$$

The estimate for the next point in the iteration comes from setting $g(x) = 0$:

$$x^{i+1} = x^i - \frac{g(x^i)}{g'(x^i)}.$$

Applying the method to the problem at hand, the Taylor polynomial for the force imbalance is written as

$$\Delta F(u) = \Delta F(u^i) + \frac{\partial \Delta F}{\partial u}\Big|_{u=u^i}(u - u^i).$$

Since the external force F is constant, the derivative simplifies as

$$\frac{\partial \Delta F}{\partial u}\Big|_{u=u^i} = -\frac{\partial F_{int}}{\partial u}\Big|_{u=u^i}.$$

Assuming that $\Delta F(u)$ is zero as above, we get the next approximate displacement solution u^{i+1} from the equation

$$\Delta F(u^i) = \frac{\partial F_{int}}{\partial u}\Big|_{u=u^i}(u^{i+1} - u^i).$$

Introducing

$$\Delta u^{i+1} = u^{i+1} - u^i$$

and comparing with the relation between the force imbalance and incremental displacement yields

$$K^i = \frac{\partial F_{int}}{\partial u}\Big|_{u=u^i}.$$

Hence, the K^i is commonly called the tangent stiffness. Finally,

$$\Delta u^{i+1} = K^{i,-1}\Delta F^i.$$

A simplified algorithm of such an iterative scheme is as follows:

For $i = 1, 2, \ldots$ until convergence, compute:

1. Internal force: $F_{int}^i = F_{int}(u = u^i)$

2. Force imbalance: $\Delta F^i = F - F_{int}^i$

3. Tangent stiffness: $K_t^i = -\frac{\partial F_{int}}{\partial u}\Big|_{u=u^i}$

4. Incremental displacement: $\Delta u^{i+1} = K^{i,-1}\Delta F^i$

5. Updated displacement: $u^{i+1} = u^i + \Delta u^{i+1}$

End loop on i.

The initial displacement u^1 and the initial internal force F_{int}^1 may or may not be zero. Note, that the initial tangent stiffness K^1 is assumed to be not zero. Figure 16.3 shows three steps of the iterative process.

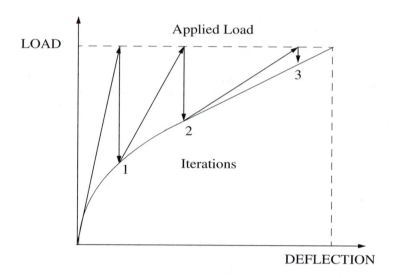

FIGURE 16.3 Newton-Raphson iteration

If the u^i displacement is close to the exact solution u^* then the convergence rate of Newton's method is quadratic, that is

$$||u^* - u^{i+1}|| \leq ||u^* - u^i||^2.$$

This may not be true for an initial displacement very far from the equilibrium or when the tangent (stiffness) changes direction. These are issues of concern in practical implementations of the method and some heuristics are required to solve them. The calculation of specific convergence criteria will be discussed in Section 16.4.

The price for this good convergence rate is the rather time-consuming step of the algorithm: the solution for the incremental displacement. In practice, the inverse of the tangent stiffness matrix is not explicitly computed, the factorization and forward-backward substitution operations shown in Chapter 7 are used. In contrast, in linear static analysis, as discussed in earlier chapters, there is only one factorization and forward backward substitution required. In nonlinear analysis many of those steps may be needed to find the equilibrium.

The computational efficiency may be improved by modifying the Newton iteration method. In this method, depicted in Figure 16.4, the tangent stiffness matrix is updated only at selected but not all the steps. This lessens the number of solutions (factorization and substitution) required. The decision whether to update the matrix may be based on the last incremental displacement.

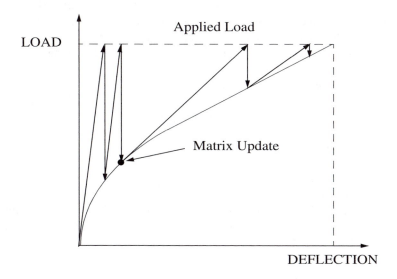

FIGURE 16.4 Modified Newton iteration

By comparing Figures 16.3 and 16.4, one can see that the price of computational efficiency is numerical accuracy. While the modified method saves time by computing fewer updates and factors, the incremental displacements are becoming smaller, resulting in a slower convergence rate. It is possible, however, to produce a both computationally and numerically efficient nonlinear iteration technique, as shown in the following section.

16.4 Quasi-Newton iteration techniques

The Newton method may be further improved by only computing an approximate update and inverse of the tangent stiffness matrix. Such a computation of the updated tangent stiffness matrix is called a quasi-Newton method and the scheme is similar to the one shown in Figure 16.4, however, the update now is a quasi-Newton update.

The quasi-Newton update calculates a secant type approximation of the tangent stiffness matrix based on the two previous iterations. One of the most frequently used methods in this class is the BFGS method, (Broyden-Fletcher-Goldfarb-Shanno [2] [3] [4] [5]) discussed in the following.

We first introduce

$$\gamma = \Delta F^{i-1} - \Delta F^i,$$

and for notational convenience

$$\delta = u^i - u^{i-1},$$

although it is really equivalent to Δu^i. The approximate updated stiffness matrix based on the last two points and the just introduced quantities is proposed as

$$K^{i+1} = K^i + \frac{\gamma \gamma^T}{\gamma^T \delta} - \frac{K^i \delta \delta^T K^i}{\delta^T K^i \delta}.$$

This is a rank two update of a matrix. The Sherman-Morrison formula enables the calculation of the inverse of an updated matrix based on the original inverse and the update information. The general formula is

$$(A + uv^T)^{-1} = A^{-1} + \frac{A^{-1} uv^T A^{-1}}{1 + v^T A^{-1} u},$$

where A is an n by n matrix updated by the u and v vectors. The formula computes the inverse of the updated matrix in terms of the inverse of the original matrix and the updating vectors.

Applying the formula for K^i being updated by two pairs of vectors simultaneously yields the BFGS formula:

$$K^{i+1,-1} = K^{i,-1} + (1 + \frac{\gamma^T K^{i,-1} \gamma}{\gamma^T \delta}) \frac{\delta \delta^T}{\delta^T \gamma} - \frac{\delta \gamma^T K^{i,-1} + K^{i,-1} \gamma \delta^T}{\gamma^T \delta}.$$

This is a rather complex formula and will be reformulated for computational purposes. Nevertheless, it directly produces the inverse for the next approximate stiffness from the last inverse and information from the last two steps.

We can reformulate the BFGS update as the sum of a triple product and a vector update as

$$K^{i,-1} = A^T K^{i-1,-1} A + \frac{\delta \delta^T}{\delta^T \gamma}.$$

Here

$$A = I - \frac{\gamma \delta^T}{\delta^T \gamma}.$$

This is a form more convenient for computer implementation as it is possible to calculate the displacement increment without explicitly computing the new inverse.

$$\Delta u^{i+1} = K^{i,-1} \Delta F^i.$$

This may be done in terms of the following intermediate computational steps. Compute scalars

$$a = (\delta^T \gamma)^{-1}$$

and

$$b = a(\delta^T \Delta F^i).$$

Compute vector

$$q = \Delta F^i - b\gamma.$$

Solve with earlier inverse

$$r = K^{i-1,-1} q.$$

Compute scalar

$$c = a(\gamma^T r).$$

The incremental displacement due to the BFGS update

$$\Delta u^{i+1} = r + (b - c)\delta.$$

This implicit BFGS update may be executed recursively.

16.5   Convergence criteria

We need to establish convergence criteria for both the incremental displacement and the load imbalance. Let us introduce the ratio of the incremental displacements

$$q = \frac{||\Delta u^{i+1}||}{||\Delta u^i||}.$$

An upper bound for the displacement error at step i may be written as

$$||u - u^i|| \leq ||u - u^{n+i+1}|| + ||u^{n+i+1} - u^{n+i}|| + \dots + ||u^{i+1} - u^i||,$$

where n is the additional number of iterations required for convergence. We assume that q is constant during the iteration. Then

$$||u - u^i|| \leq ||\Delta u^i||(q^n + q^{n-1} + \dots + q).$$

Finally, with the assumption that q is less than one, taking the limit of n to infinity and summing the geometric series we get

$$||u - u^i|| \leq ||\Delta u^i|| \frac{q}{1 - q}.$$

This convergence criterion may not be very accurate due to the assumptions. To represent fluctuations in the incremental displacement ratio, a form of

$$q^i = c \frac{||\Delta u^i||}{||\Delta u^{i-1}||} + (1 - c)q^{i-1}$$

is sometimes used in the practice, with c being a constant less then unity. An error function of the displacements can be formulated as

$$\epsilon_u^i = \frac{||\Delta u^i||}{||u^i||} \frac{q^i}{1 - q^i}.$$

Finally, the error criterion for the load imbalance is composed as

$$\epsilon_F^i = \frac{||\Delta F^i * u^i||}{||(F + F_{int}^i) * u^i||},$$

where the $*$ implies term by term multiplication of the vectors.

Convergence is achieved when either the condition for the load imbalance:

$$\epsilon_F^i \leq \epsilon_F,$$

or for the incremental displacement:

$$\epsilon_u^i \leq \epsilon_u$$

is satisfied. The ϵ_F, ϵ_u values are set by the engineer.

16.6   Computational example

Let us demonstrate the above formulations with a simple computational example. We will consider a single degree of freedom system with the following input:

External load: $F = 3$.

Initial displacement: $u^1 = 1$.

Let the characteristics of the system described by the internal load of

$$F_{int} = 1 + \sqrt{u}.$$

Using the relationship developed above, the tangent stiffness is then

$$K = \frac{1}{2\sqrt{u}}.$$

The theoretical equilibrium is at $u = 4$ when $F_{int} = 1 + \sqrt{4} = 3 = F$.

Let us first consider two Newton steps.

$i = 1$

$$\Delta F^1 = F - F_{int}(u = u^1) = 3 - (1 + \sqrt{1}) = 1$$

$$K^1 = \frac{1}{2\sqrt{u^1}} = \frac{1}{2}, \quad K^{1,-1} = 2$$

$$\Delta u^2 = K^{1,-1}\Delta F^1 = 2 * 1 = 2$$

$$u^2 = u^1 + \Delta u^2 = 1 + 2 = 3$$

$i = 2$

$$\Delta F^2 = F - F_{int}(u = u^2) = 3 - (1 + \sqrt{3}) = 2 - \sqrt{3}$$

$$K^2 = \frac{1}{2\sqrt{u^2}} = \frac{1}{2\sqrt{3}}, \quad K^{2,-1} = 2\sqrt{3}$$

$$\Delta u^3 = K^{2,-1}\Delta F^2 = 2\sqrt{3}(2 - \sqrt{3}) = 4\sqrt{3} - 6$$

$$u^3 = u^2 + \Delta u^3 = 3 + 4\sqrt{3} - 6 = 3.9282$$

That is now very close to the theoretical solution.

Now we replace the last step with a modified Newton step as follows.

$i = 2$

$$\Delta F^2 = F - F_{int}(u = u^2) = 3 - (1 + \sqrt{3}) = 2 - \sqrt{3}$$
$$\Delta u^3 = K^{1,-1}\Delta F^2 = 2(2 - \sqrt{3}) = 4 - 2\sqrt{3}$$
$$u^3 = u^2 + \Delta u^3 = 3 + 4 - 2\sqrt{3} = 3.5359$$

This value is clearly not as close to the theoretical solution as the regular Newton step, however, as the tangent stiffness was not updated and inverted, at a much cheaper computational price.

Finally, we execute the step with BFGS update.

$i = 2$

$$\delta = u^2 - u^1 = 3 - 1 = 2$$
$$\gamma = \Delta F^1 - \Delta F^2 = 1 - (2 - \sqrt{3}) = \sqrt{3} - 1$$
$$\overline{K}^{2,-1} = \frac{\delta\delta^T}{\delta^T\gamma} = \frac{\delta}{\gamma} = \frac{2}{\sqrt{3} - 1}$$
$$\Delta u^3 = \overline{K}^{2,-1}\Delta F^2 = \frac{2}{\sqrt{3} - 1}(2 - \sqrt{3}) = 0.7321$$
$$u^3 = u^2 + \Delta u^3 = 3 + 0.7321 = 3.7321$$

This result demonstrates the power of the method, without explicitly updating the stiffness matrix ($\overline{K}$ indicates the approximate inverse), we obtained a solution better than that of the modified method. Note, that the update is in the simplified form since

$$A = I - \frac{\gamma\delta^T}{\delta^T\gamma} = 0$$

for the single degree of freedom (scalar) problem.

It is also notable that the approximate stiffness (not explicitly computed) would be

$$\overline{K}^2 = \frac{\sqrt{3} - 1}{2}.$$

On the other hand, the slope of the secant between the points (u^1, F_{int}^1) and (u^2, F_{int}^2) is the same value. As mentioned above the quasi-Newton method is a secant type approximation.

16.7 Nonlinear dynamics

The solution of nonlinear transient (or frequency) response problems introduces additional complexities. The solution is still based on time integration
schemes as shown in Chapter 14, however, that is now embedded into the
nonlinear iterations. The governing equation in this case is:

$$M\ddot{u}(t) + B\dot{u}(t) + Ku(t) = F(t) - F_{int}(u(t)).$$

Note, that the external load is considered to have time-dependence, but not
dependent on the displacement. The Newmark β method introduced in Chapter 14 may also be recast in a two point recurrence formula more suitable for
the nonlinear dynamics case:

$$u(t + \Delta t) = u(t) + \Delta t \dot{u}(t) + \frac{1}{2}\Delta t^2 \ddot{u}(t) + \beta \Delta t^2 (\ddot{u}(t + \Delta t) - \ddot{u}(t)),$$

and

$$\dot{u}(t + \Delta t) = \dot{u}(t) + \Delta t \ddot{u}(t) + \frac{1}{2}\Delta t(\ddot{u}(t + \Delta t) - \ddot{u}(t)).$$

Solving these equations for the velocity $\dot{u}(t + \Delta t)$ and acceleration $\ddot{u}(t + \Delta t)$
in terms of the displacements results in

$$\dot{u}(t + \Delta t) = \frac{1}{2\beta \Delta t}(u(t + \Delta t) - u(t)) + (1 - \frac{1}{2\beta})\dot{u}(t) + (1 - \frac{1}{4\beta})\Delta t \ddot{u}(t),$$

and

$$\ddot{u}(t + \Delta t) = \frac{1}{\beta \Delta t^2}(u(t + \Delta t) - u(t)) - \frac{1}{\beta \Delta t}\dot{u}(t) - (\frac{1}{4\beta} - 1)\ddot{u}(t).$$

Substituting these into the governing equation at the next time step considering also the iteration steps, yields

$$M\ddot{u}^{i+1}(t + \Delta t) + B\dot{u}^{i+1}(t + \Delta t) + K^i(t + \Delta t)\Delta u^{i+1} = F(t + \Delta t) - F_{int}^i(t + \Delta t),$$

where

$$K^i(t + \Delta t) = \frac{\partial F_{int}^i}{\partial u}(t + \Delta t).$$

The resulting equation is the nonlinear Newton-Raphson iteration scheme

$$[\frac{1}{\beta \Delta t^2}M + \frac{1}{2\beta \Delta t}B + K^i(t + \Delta t)]\Delta u^{i+1} = \Delta \overline{F}^i(t + \Delta t),$$

where

$$u^{i+1}(t + \Delta t) = u^i(t + \Delta t) + \Delta u^{i+1}.$$

The dynamic load imbalance is

$$\Delta \overline{F}^i(t + \Delta t) = \Delta F^i(t + \Delta t) -$$

$$\frac{M}{\beta \Delta t^2}(u^i(t + \Delta t) - u(t) - \Delta t \dot{u}(t)) + (\frac{1}{2\beta} - 1)M\ddot{u}(t)$$

$$-\frac{1}{2\beta \Delta t}B(u^i(t + \Delta t) - u(t)) + (\frac{1}{2\beta} - 1)B\dot{u}(t) - (1 - \frac{1}{4\beta})\Delta t B \ddot{u}(t).$$

Several comments are in order. First, the tangent stiffness matrix $K^i(t+\Delta t)$ may be replaced by $K(t)$ resulting in the modified Newton-Raphson scheme. Also note, that $u(t)$ is the converged displacement in the last time step and $u^i(t + \Delta t)$ is the displacement in the i-th iteration for the next time step. Finally, the tangent stiffness matrix $K^i(t+\Delta t)$ may also be replaced by $\overline{K}(t)$, representing a BFGS update.

The iteration at a time step proceeds until the force imbalance is sufficiently small. At this point the resulting displacement becomes the starting point for the next time step. Algorithmic implementation of above is rather delicate and commercial finite element analysis systems have various proprietary adjustments to control the efficiency and numerical convergence of the process.

References

[1] Bathe, K.-J.; Finite element procedures in engineering analysis, Prentice-Hall, Englewood Cliffs, New Jersey, 1988

[2] Broyden, C. G.; The convergence of a class of double rank minimization algorithms, J. Inst. Math. Appl., Vol. 6, pp. 76-90, 222-231, 1970

[3] Fletcher, R.; A new approach to variable metric algorithms, Computer Journal, Vol. 13, pp. 317-322, 1970

[4] Goldfarb, D.; A family of variable metric methods derived by variational means, Math. Comp., Vol. 24, pp. 23-26, 1970

[5] Shanno, D. F.; Conditioning of quasi-Newton methods for function minimization, Math. Comp., Vol. 24, pp. 647-656, 1970

17

Sensitivity and Optimization

The subject of this chapter is a class of computations that characterize the stability and provide the methodology for optimization of the engineering design.

17.1   Design sensitivity

At this stage of the engineering solutions we are getting close to the original engineering problem defined by the engineer. Sensitivity computations are based on computing derivatives of some computational results with respect to some changes in the model. The changes in the model are described by design variables. Design variables may be some geometric measures of the model, for example the length and width of a component. They may also be other engineering quantities, such as eigenfrequencies.

We contain the set of design variables in an array

$$x = \begin{bmatrix} x_1 \\ .. \\ x_i \\ .. \\ x_m \end{bmatrix}.$$

The design variables are related to the displacements:

$$x = x(u),$$

and conversely the displacement vector may be expressed in terms of the design variables

$$u = u(x).$$

Let us consider, again, the linear static solution

$$Ku = F.$$

The first variation of the linear static solution,

$$K\frac{\partial u}{\partial x_i} + \frac{\partial K}{\partial x_i}u = \frac{\partial F}{\partial x_i},$$

allows us to compute the sensitivity of the linear static solution with respect to the ith design variable as

$$\frac{\partial u}{\partial x_i} = K^{-1}(\frac{\partial F}{\partial x_i} - \frac{\partial K}{\partial x_i}u).$$

Assuming that the initial design variable vector is x^0, the terms on the right-hand side may be computed by finite differences as

$$\frac{\partial F}{\partial x_i} = \frac{F(x^0 + \Delta x_i) - F(x^0 - \Delta x_i)}{2\Delta x_i},$$

and

$$\frac{\partial K}{\partial x_i} = \frac{K(x^0 + \Delta x_i) - K(x^0 - \Delta x_i)}{2\Delta x_i}.$$

Hence, the sensitivity of the solution with respect to changes in the ith design variable is obtained.

17.2   Design optimization

The design variables may be automatically changed to optimize some design objective under some constraints, a process called design optimization. Let us consider, for example, a maximum stress constraint applied to the linear statics problem

$$\sigma \le \sigma_{max},$$

or written in a constraint equation form

$$g(x) = \sigma - \sigma_{max} \le 0.$$

Here the stress may be a compound measure, such as the von Mises stress described in Chapter 18 or could be a component. In the latter case multiple constraint equations are given. Other kinds of constraints may be for example the physical limits of some design variables.

The design objective may be formed as

$$f(x) = minimum.$$

For example, a design objective may be to find the minimum weight (or volume) of the model that satisfies the stress constraint. Note, that both the objective function $f(x)$ and the constraint function $g(x)$ are vector valued functions.

Finding the optimal design is based on the approximation of both the objective function

$$f(x^0 + \Delta x) = f(x^0) + \nabla f|_{x^0} \Delta x,$$

and the constraint function

$$g(x^0 + \Delta x) = g(x^0) + \nabla g|_{x^0} \Delta x.$$

The sensitivity of the objective function is computed by finite differences

$$\nabla f(x) = \begin{bmatrix} \frac{\partial f}{\partial x_1} \\ .. \\ \frac{\partial f}{\partial x_i} \\ .. \\ \frac{\partial f}{\partial x_m} \end{bmatrix} = \begin{bmatrix} \frac{f(x^0 + \Delta x_1) - f(x^0)}{\Delta x_1} \\ .. \\ \frac{f(x^0 + \Delta x_i) - f(x^0)}{\Delta x_i} \\ .. \\ \frac{f(x^0 + \Delta x_m) - f(x^0)}{\Delta x_m} \end{bmatrix}.$$

The constraint sensitivities are

$$\nabla g(x) = \begin{bmatrix} \frac{\partial g}{\partial x_1} \\ .. \\ \frac{\partial g}{\partial x_i} \\ .. \\ \frac{\partial g}{\partial x_m} \end{bmatrix}.$$

The derivatives are

$$\frac{\partial g}{\partial x_i} = \frac{\partial g}{\partial \sigma} \frac{\partial \sigma}{\partial x_i}.$$

As we introduced earlier the stress vector is related to the displacements as

$$\sigma = DBu,$$

hence

$$\frac{\partial \sigma}{\partial x_i} = \frac{\partial \sigma}{\partial u} \frac{\partial u}{\partial x_i} = DB \frac{\partial u}{\partial x_i}.$$

Here $\frac{\partial u}{\partial x_i}$ is the sensitivity we computed in the prior section. With that

$$\frac{\partial g}{\partial x_i} = DB \frac{\partial u}{\partial x_i}.$$

To minimize the objective function, one direction in which the model may be suitably modified is the direction opposite to the gradient

$$d = -\nabla f.$$

This is the method called the steepest descent. However, this direction may violate the constraints so it may need to be modified. The feasible directions are mathematically described as

$$(\nabla f(x))^T d \leq 0.$$

The equality in this equation results in the direction vector of the steepest descent. Use of any other directions results in the feasible direction method. The range of the directions is from 90 to 270 degrees from the gradient of the objective function.

To avoid violating the constraints, the

$$(\nabla g(x))^T d \leq 0$$

condition must also be satisfied. The range of this direction vector is again from 90 to 270 degrees from the gradient of the constraint equation. The direction that satisfies both may be found by various heuristic techniques, such as the modified feasible direction method [7] frequently used in the industry. The formulation also enables the application of advanced automated mathematical optimization techniques, such as linear or nonlinear programming.

The modification of the model after a step of this optimization is

$$x^1 = x^0 + \alpha d,$$

where α is a scalar (distance) variable and d is the direction chosen to obey all conditions. This is a very powerful approach since with the help of this single scalar we can modify the value of a large number of design variables.

The objective function and the constraint equations are adjusted accordingly:

$$f^1 = f(x^1),$$

and

$$g^1 = g(x^1).$$

It is assumed that

$$f^1 \leq f^0,$$

in the sense the minimum is defined. From this new design variable set we can again compute the sensitivities and the process is repeated until the constrained optimum is achieved. At the constrained optimum the Kuhn-Tucker condition applies. The condition states that at the constrained optimum the vector sum of the objective and constraint gradients must be zero with some appropriately chosen multiplying factors. Mathematically

$$\lambda \nabla g(x^*) + \nabla f(x^*) = 0,$$

where the x^* is the optimal design variable setting. The design cannot be improved further. The λ is the appropriate (Lagrange) multiplier. Note again, that most of the time there are several constraint conditions given resulting in multiple ∇g vectors in the condition.

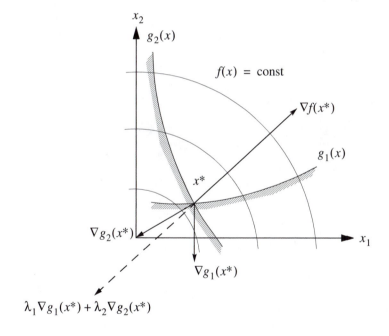

FIGURE 17.1 Optimum condition

The graphical representation of the condition is shown in Figure 17.1 with two constraint conditions.

17.3   Planar bending of the bar

During the course of the book we have discussed the bar element several times
in order to present the computational material that was our main subject at
that particular point. We started with a rigid bar in connection with multi-
point constraints. Then the bar was allowed to have axial flexibility and later
it was allowed to undergo large rotations. Now, as a foundation for an optimization example, we consider an in-plane bending of the bar.

Let us consider the bar shown in Figure 17.2 and assume that it is bending
in the x, y plane. Ignore the fact of being supported at the left end for the
moment. The bending depends on the forces and moments applied at the
ends.

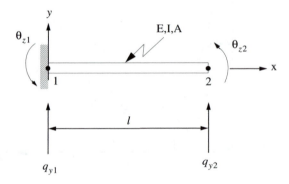

FIGURE 17.2 Planar bending of bar element

The possible deformations of the planar bending of the bar element are the
vertical displacements (q_y) and rotations around the z axis (θ_z) for both of
the two node points.

$$q_e = \begin{bmatrix} q_{y1} \\ \theta_{z1} \\ q_{y2} \\ \theta_{z2} \end{bmatrix}.$$

Let us assume that the vertical displacement function is a cubic polynomial. This is a prudent choice based on the following argument. The mathematical model describing the bar bending problem is

$$EI\frac{d^4 q_y}{dx^4} = F(x),$$

where the I is the moment of inertia with respect to the bending axis of the cross section (the z axis) and E is the already introduced Young's modulus [1]. The right-hand side is the load vector, as function of x it is distributed over the length of the bar. It may also be a point load or a moment applied at a certain point of the bar.

The homogeneous differential equation has a solution form:

$$q_y(x) = a_0 + a_1 x + a_2 x^2 + a_3 x^3,$$

hence above assumption. Then the following is true:

$$q_{y1} = q_y(0) = a_0,$$

and

$$q_{y2} = q_y(l) = a_0 + a_1 l + a_2 l^2 + a_3 l^3.$$

The rotation angle and the slope of the vertical displacement are related as

$$\theta_z = -\frac{dq_y}{dx}.$$

Utilizing this results in

$$\theta_{y1} = -a_1,$$

and

$$\theta_{y2} = -a_1 - 2a_2 l - 3a_3 l^2.$$

We are now seeking a matrix of the shape functions to describe the vertical displacement field

$$q = N q_e,$$

where

$$N = \begin{bmatrix} N_1 & N_2 & N_3 & N_4 \end{bmatrix}.$$

Some tedious algebra produces the satisfactory shape functions as

$$N_1 = 1 - 3\frac{x^2}{l^2} + 2\frac{x^3}{l^3},$$

$$N_2 = -x - 2\frac{x^2}{l} - \frac{x^3}{l^2},$$

$$N_3 = 3\frac{x^2}{l^2} - 2\frac{x^3}{l^3},$$

and

$$N_4 = -\frac{x^2}{l} + \frac{x^3}{l^2}.$$

The strain due to the bending of the bar is

$$\epsilon = -y\frac{d^2 q_y}{dx^2}.$$

That is distinctly different from the axial strain of the bar under tension or compression as shown in Section 4.3. The strain-displacement relationship of

$$\epsilon = Bq_e$$

is satisfied with the matrix

$$B^T = \begin{bmatrix} -\frac{6}{l^2} + \frac{12x}{l^3} \\ -\frac{6x}{l^2} + \frac{4}{l} \\ \frac{6}{l^2} - \frac{12x}{l^3} \\ \frac{2}{l} - \frac{6x}{l^2} \end{bmatrix},$$

where the algebraic details are again omitted. Note, that due to the form of the shape functions, the B matrix is not constant. The stiffness matrix for the bending bar element finally is

$$k_e = \int_{x=0}^{x=l} B^T D B dx = EI \int_{x=0}^{x=l} B^T B dx,$$

with the stress-strain matrix for bending being

$$D = EI.$$

Executing the multiplication and integration results

$$k_e = EI \begin{bmatrix} \frac{12}{l^3} & \frac{6}{l^2} & -\frac{12}{l^3} & \frac{6}{l^2} \\ \frac{6}{l^2} & \frac{4}{l} & -\frac{6}{l^2} & \frac{2}{l} \\ -\frac{12}{l^3} & -\frac{6}{l^2} & \frac{12}{l^3} & -\frac{6}{l^2} \\ \frac{6}{l^2} & \frac{2}{l} & -\frac{6}{l^2} & \frac{4}{l} \end{bmatrix}.$$

Note, that this is the stiffness matrix for the bar bending in the x, y plane only. One can, however, easily modify this equation for bending in another plane by systematic changes. General formulation for simultaneous deformations in both planes and other variations are also possible [3] but beyond our focus.

The finite element equilibrium of the cantilever bar problem of Figure 17.2, modeled by a single bar element is now written:

$$k_e q = f,$$

or

$$k_e \begin{bmatrix} q_{y1} \\ \theta_{z1} \\ q_{y2} \\ \theta_{z2} \end{bmatrix} = \begin{bmatrix} F_{y1} \\ M_{z1} \\ F_{y2} \\ M_{z2} \end{bmatrix},$$

where F are point loads and M are moments applied at the ends. Let us now constrain the left end and apply only a point load at the right end, $F_{y2} = F$. Then

$$k_e \begin{bmatrix} 0 \\ 0 \\ q_{y2} \\ \theta_{z2} \end{bmatrix} = \begin{bmatrix} 0 \\ 0 \\ F \\ 0 \end{bmatrix}.$$

The lower 2×2 partition of the equilibrium yields

$$EI \begin{bmatrix} \frac{12}{l^3} & -\frac{6}{l^2} \\ -\frac{6}{l^2} & \frac{4}{l} \end{bmatrix} \begin{bmatrix} q_{y2} \\ \theta_{z2} \end{bmatrix} = \begin{bmatrix} F \\ 0 \end{bmatrix}.$$

Inverting gives the solution of

$$\begin{bmatrix} q_{y2} \\ \theta_{z2} \end{bmatrix} = \frac{1}{EI} \begin{bmatrix} \frac{l^3}{3} & \frac{l^2}{2} \\ \frac{l^2}{2} & l \end{bmatrix} \begin{bmatrix} F \\ 0 \end{bmatrix} = \begin{bmatrix} \frac{Fl^3}{3EI} \\ \frac{Fl^2}{2EI} \end{bmatrix}.$$

Finally, based on the vertical displacement and the deflection angle solution the maximum stress of the bar may be obtained as

$$\sigma_{max} = E\epsilon = EBq_e = \frac{Fly}{2I},$$

where y is the vertical measure of the cross section profile of the bar.

17.4 Computational example

We consider the example of a bar constrained at the left end (see Figure 17.2) and loaded with a constant force F on the right end. The objective is to optimize the shape of the cross section of the bar to minimize the material needed while subject to the constraint of a given maximum stress σ_{max}.

Let the bar have a rectangular cross section of width x_1 and height x_2. These two will be chosen as the design variables, as hinted by the notation also.

The maximum stress of such bar with length l based on the results of the last section is

$$\sigma_{max} = \frac{6Fl}{x_1 x_2^2},$$

with substituting the cross section moment of inertia.

This formula would of course be not explicitly available for a multiple element model; the stresses would have to be evaluated from the displacements as shown in Section 18.2 in more detail. This fact does not take away from the generality of the following example, whose goal is to demonstrate some of the steps of optimization computations.

The volume of the bar is

$$V = x_1 x_2 l.$$

For the sake of simplicity of the example, we consider a unit length bar, $l = 1$ and the loading force of $F = \frac{1}{6}$.

From engineering practicality it follows that the shape of the cross section is bounded, neither a very thin but wide nor a very narrow but thick cross section is desirable. These may be expressed as

$$\frac{x_2}{x_1} \leq r_{max},$$

and

$$\frac{x_2}{x_1} \geq r_{min}.$$

Again for simplicity's sake we will use

$$r_{max} = 2 \quad r_{min} = \frac{1}{2}$$

in the following. These are quite practical limits, a cross section of height twice the width, or half the width are frequently used manufactured profiles.

The optimization problem is now formally posed as follows.

In the design space of

$$x = \begin{bmatrix} x_1 \\ x_2 \end{bmatrix},$$

minimize the volume, i.e. the cross section area:

$$f(x) = x_1 x_2$$

subject to the stress constraint of

$$g_1(x) = \frac{1}{x_1 x_2^2} - 1 \leq 0$$

and shape constraints of

$$g_2(x) = \frac{1}{2} - \frac{x_2}{x_1} \leq 0$$

and

$$g_3(x) = \frac{x_2}{x_1} - 2 \leq 0.$$

Here we used a unit numeric limit for the stress maximum

$$\sigma_{max} = 1.$$

The two-dimensional design space of the two variables is graphically shown on Figure 17.3.

The level curves of

$$x_2 = \frac{c}{x_1}$$

demonstrate the objective function's behavior. The inequalities of

$$x_2 \leq 2x_1$$

and

$$x_2 \geq \frac{1}{2} x_1$$

represent the boundaries of admissible cross section shapes. Finally, the inequality

$$x_2 \geq \sqrt{\frac{1}{x_1}}$$

defines the boundary of the feasible region with respect to the stress constraint. It is clear from the arrangement that the objective function is decreasing towards the origin. The objective function's gradient may be calculated as

$$\nabla f(x) = \begin{bmatrix} \frac{\partial f}{\partial x_1} \\ \frac{\partial f}{\partial x_2} \end{bmatrix} = \begin{bmatrix} x_2 \\ x_1 \end{bmatrix}.$$

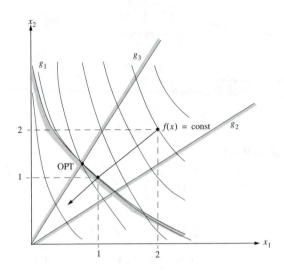

FIGURE 17.3 Design space of optimization example

The feasible direction with respect to the objective function is simply

$$d = -x_2 i - x_1 j$$

pointing back toward the origin. Let us assume a quite feasible starting point of the optimization at $x_1 = 2, x_2 = 2$, a square cross-section. The corresponding feasible direction vector is

$$d = -2i - 2j.$$

This vector will clearly intersect the stress constraint curve $g_1(x)$ at the point $x_1 = 1, x_2 = 1$, so we could consider this the next point in the optimization sequence. The feasible direction vector is now

$$d = -i - j,$$

which needs to be modified as it immediately violates the stress constraint.

A modification of the feasible direction vector to the left and to the right is possible until reaching the intersection of the shape constraint lines and the stress constraint curve. From these points the closer to the x_2 axis

$$x_1 = \frac{1}{(4)^{\frac{1}{3}}},$$

$$x_2 = \frac{2}{(4)^{\frac{1}{3}}}$$

lies at lower level of the objective function. In order to verify that we have reached an optimum, we will use the Kuhn-Tucker condition. The gradient of the upper shape constraint is

$$\nabla g_3(x) = \begin{bmatrix} \frac{\partial g_3}{\partial x_1} \\ \frac{\partial g_3}{\partial x_2} \end{bmatrix} = \begin{bmatrix} \frac{-x_2}{x_1^2} \\ \frac{1}{x_1} \end{bmatrix} = -2(4)^{\frac{1}{3}} i + (4)^{\frac{1}{3}} j.$$

The gradient of the stress constraint is

$$\nabla g_1(x) = \begin{bmatrix} \frac{\partial g_1}{\partial x_1} \\ \frac{\partial g_1}{\partial x_2} \end{bmatrix} = \begin{bmatrix} \frac{-1}{x_1^2 x_2} \\ \frac{-2}{x_1 x_2^3} \end{bmatrix} = -(4)^{\frac{1}{3}} i - (4)^{\frac{1}{3}} j.$$

Finally, the gradient of the objective function at this point is

$$\nabla f(x) = \frac{2}{(4)^{\frac{1}{3}}} i + \frac{1}{(4)^{\frac{1}{3}}} j.$$

If this point is an optimum, the Kuhn-Tucker condition would be

$$\nabla f(x) + \lambda_1 \nabla g_1 x + \lambda_2 \nabla g_2 x = 0.$$

With the selection of

$$\lambda_1 = \frac{1}{3(4)^{\frac{2}{3}}}$$

and

$$\lambda_2 = \frac{4}{3(4)^{\frac{2}{3}}}$$

the condition is satisfied and an optimum exist at that point.

The optimal cross section shape has width$= \frac{1}{(4)^{\frac{1}{3}}}$ and height$= \frac{2}{(4)^{\frac{1}{3}}}$ units resulting in the minimum volume of

$$V = \frac{2}{(4)^{\frac{2}{3}}} = \frac{1}{(2)^{\frac{1}{3}}}$$

cubic units. Considering the starting material volume of 4 units, the reduction is rather significant.

In practice, the resulting sizes may not be completely appropriate for measurement and manufacturing purposes. Therefore, the resulting dimensions are likely to be rounded up to the nearest measurable level. For our example the final results of width = 0.63 and height = 1.26 are practical. This slightly increases the theoretical optimum volume to the practical 0.794 which is still a significant improvement.

In the preceding, emphasis was given to the fact that the procedure outlined finds "an" optimum, not necessarily "the" optimum. This is an important aspect of such techniques: a local optimum in the neighborhood is found, but the global optimum may not be found.

The heuristic procedure for modifying the search direction, on the other hand, has not been exposed in very much detail beyond the conceptual level. Such algorithms are often proprietary and the actual implementation in a software environment contributes to the efficiency as much as the cleverness of the algorithm.

17.5   Eigenfunction sensitivities

This topic has been subject of the interest in the area of perturbation theory for linear differential operators literally for centuries. For example, [4] has already presented results for eigenvalue derivatives almost 160 years ago.

The sensitivities of the linear eigenvalue problem

$$(K_{aa} - \lambda^i M_{aa})\phi_a^i = 0$$

deserve further consideration. Here λ^i, ϕ_a^i is the ith eigenpair of the problem, $i = 1, m$. Differentiation with respect to the jth design variable and reordering results in

$$(K_{aa} - \lambda^i M_{aa})\frac{\partial \phi_a^i}{\partial x_j} + \left(\frac{\partial K_{aa}}{\partial x_j} - \lambda^i \frac{\partial M_{aa}}{\partial x_j}\right)\phi_a^i = \frac{\partial \lambda^i}{\partial x_j} M_{aa}\phi_a^i.$$

Pre-multiplying the equation with $\phi_a^{i,T}$ results in the sensitivity of the ith eigenvalue with respect to the jth design variable yields

$$\frac{\partial \lambda^i}{\partial x_j} = \frac{\phi_a^{i,T}\left(\frac{\partial K_{aa}}{\partial x_j} - \lambda^i \frac{\partial M_{aa}}{\partial x_j}\right)\phi_a^i}{\phi_a^{i,T} M_{aa}\phi_a^i},$$

as the first term on the left-hand side vanishes. The derivative of the stiffness matrix was computed by finite differences in the prior section. Similarly, the mass matrix derivative is

$$\frac{\partial M_{aa}}{\partial x_i} = \frac{M_{aa}(x^0 + \Delta x_i) - M_{aa}(x^0 - \Delta x_i)}{2\Delta x_i}.$$

Note, that this equation is valid only for simple eigenvalues.

Another reorganization yields the equation for the eigenvector sensitivity

$$(K_{aa} - \lambda^i M_{aa})\frac{\partial \phi_a^i}{\partial x_j} = (\frac{\partial \lambda^i}{\partial x_j} M_{aa} + \lambda^i \frac{\partial M_{aa}}{\partial x_j} - \frac{\partial K_{aa}}{\partial x_j})\phi_a^i.$$

This form contains the eigenvalue sensitivity and the mass and stiffness matrix derivatives which may not be available. The following approximate form for the eigenvector sensitivity is based on [6]. Let us assume a slight perturbation of the jth design variable with the amount of Δx_j. Then the right-hand side of above equation may be approximated by

$$(K_{aa} - \lambda M_{aa})\frac{\partial \phi_a^i}{\partial x_j} = \frac{1}{\Delta x_j}(\Delta \lambda^i M_{aa} + \lambda^i \Delta M_{aa} - \Delta K_{aa})\phi_a^i.$$

Here

$$\Delta K_{aa} = K_{aa}|_{x^j + \Delta x^j} - K_{aa}|_{x^j},$$

and

$$\Delta M_{aa} = M_{aa}|_{x^j + \Delta x^j} - M_{aa}|_{x^j}.$$

With the ith modal mass

$$m^i = \phi_a^{i,T} M_{aa} \phi_a^i,$$

the third approximation component of the right-hand side is

$$\Delta \lambda^i = \frac{1}{m^i}(\phi_a^{i,T} \Delta K_{aa} \phi_a^i - \lambda^i \phi_a^{i,T} \Delta M_{aa} \phi_a^i).$$

With these, the complete right-hand side is

$$R_{aa}^i = \frac{1}{\Delta x_j}(\Delta \lambda^i M_{aa} + \lambda^i \Delta M_{aa} - \Delta K_{aa})\phi_a^i.$$

Finally, denoting

$$A_{aa}^i = K_{aa} - \lambda^i M_{aa},$$

we can present the problem as a system of linear equations

$$A_{aa}^i \frac{\partial \phi_a^i}{\partial x_j} = R_{aa}^i.$$

By definition the matrix of the system of n linear equations is singular. The rank deficiency is, however, only one which may be overcome by arbitrarily setting one component of the solution vector. If we set it to zero and eliminate the corresponding row and column of A^i_{aa} the problem may be solved and the eigenvector sensitivities obtained.

There are ways to calculate the eigenfunction derivatives analytically [5]. These methods are not attractive to the very large problems occurring in the industry and as such are not widely used in the practice.

17.6   Variational analysis

The section addresses the issue of calculating the variation of the response of a structure with respect to a variation in a design parameter.

Let us consider the linear statics problem

$$Ku = F,$$

where K is the stiffness matrix and the F is the load vector. The stiffness matrix is assembled as

$$K = \Sigma^{n_e}_{e=1} K_e,$$

where K_e is the eth element matrix.

For the sake of simplicity, consider a single, scalar design variable, however, the forthcoming is true for a set of design variables also. We assume that both K and F are functions of the design variable. The goal of the development is to establish the solution u as a function of the design variable x.

In order to achieve that, express the solution in a Taylor series

$$u(x) = u_0 + \frac{\partial u}{\partial x} x + \frac{\partial^2 u}{\partial x^2} \frac{x^2}{2} + \dots.$$

Here u_0 is the fixed static solution

$$u_0 = K^{-1} F,$$

computed via the factorization

$$K = LDL^T.$$

Considering n terms in the series we get

$$u = \Sigma_{j=0}^n \frac{\partial^j u}{\partial x^j} \frac{x^j}{j!}.$$

A similar series expression is assumed for the element matrices as

$$K_e = \Sigma_{j=0}^n \frac{\partial^j K_e}{\partial x^j} \frac{x^j}{j!}.$$

Let us now consider derivatives of the equilibrium equation with respect to the design variable. The kth derivative is

$$\frac{\partial^k K}{\partial x^k} u + K \frac{\partial^k u}{\partial x^k} = \frac{\partial^k F}{\partial x^k}.$$

Substituting the element matrices

$$\Sigma_{e=1}^{n_e} \frac{\partial^k K_e}{\partial x^k} u + K \frac{\partial^k u}{\partial x^k} = \frac{\partial^k F}{\partial x^k}.$$

The kth derivative of an element matrix may be written as

$$\frac{\partial^k K_e}{\partial x^k} = \Sigma_{j=0}^{k-1} a_j \frac{\partial^j K_e}{\partial x^j},$$

where

$$a_j = \frac{x^j}{j!}.$$

Substituting results in

$$(\Sigma_{e=1}^{n_e} \Sigma_{j=0}^{k-1} a_j \frac{\partial^j K_e}{\partial x^j}) u + K \frac{\partial^k u}{\partial x^k} = \frac{\partial^k F}{\partial x^k}.$$

Considering that u is also a polynomial expression in a_j albeit with different coefficients, truncating it at the $(k-1)$st term we may write

$$(\Sigma_{e=1}^{n_e} \Sigma_{j=0}^{k-1} a_j \frac{\partial^j K_e}{\partial x^j}) u = \Sigma_{e=1}^{n_e} \Sigma_{j=0}^{k-1} b_j \frac{\partial^{k-1-j} K_e}{\partial x^{k-1-j}} \frac{\partial^j u}{\partial x^j},$$

where the binomial coefficients are

$$b_j = \frac{(k-1)!}{(k-1-j)! j!}.$$

Reorganization yields a recursive equation for computing the derivatives of the displacement with respect to the design variable.

$$K \frac{\partial^k u}{\partial x^k} = \frac{\partial^k F}{\partial x^k} - \Sigma_{e=1}^{n_e} \Sigma_{j=0}^{k-1} b_j \frac{\partial^{k-1-j} K_e}{\partial x^{k-1-j}} \frac{\partial^j u}{\partial x^j}.$$

Specifically, for $k = 1$,

$$K \frac{\partial u}{\partial x} = \frac{\partial F}{\partial x} - \Sigma_{e=1}^{n_e} K_e u_0$$

gives the first derivative. The second derivative is obtained from

$$K \frac{\partial^2 u}{\partial x^2} = \frac{\partial^2 F}{\partial x^2} - \Sigma_{e=1}^{n_e} \Sigma_{j=0}^{1} b_j \frac{\partial^{1-j} K_e}{\partial x^{1-j}} \frac{\partial^j u}{\partial x^j} = \frac{\partial^2 F}{\partial x^2} - \Sigma_{e=1}^{n_e} (K_e u_0 + \frac{\partial K_e}{\partial x} \frac{\partial u}{\partial x}),$$

and recursively on for higher derivatives. The computational complexity of the approach lies in the fact that the element stiffness matrices and the derivatives have to be reevaluated at all settings of the design variable, while the factorization of K is available from the initial static solution.

The cost of the evaluation of the right-hand side for each derivative may be lessened considering the fact that the element matrices are assembled as

$$K_e = \Sigma_{i=1}^{n_g} J_i w_i B^T D B,$$

where n_g is the number of Gauss points in the element. The J_i Jacobian evaluated at a Gauss point, w_i Gaussian weights and the D material constitutive matrix do not change. The B strain-displacement matrix needs to be modified for every solution step.

In practice, the number of derivatives evaluated should be rather small to prevent this cost becoming prohibitive. With two derivatives, for example, the variational solution of the linear statics problem is represented by

$$u(x) = u(x_0) + \frac{\partial u}{\partial x} x + \frac{\partial^2 u}{\partial x^2} \frac{x^2}{2},$$

where

$$u(x_0) = u_0.$$

The usage of three design variable values, one below and one above a nominal value (x_0), has the merit of representing the range of a variational design variable. Let these be denoted by

$$x_0 - \Delta x, \ x_0, \ x_0 + \Delta x.$$

This situation occurs, for example, when taking a symmetric manufacturing tolerance of a component into consideration. Such an approach provides an easy evaluation of the yet unresolved derivative of the load vector via finite differences:

$$\frac{\partial F}{\partial x} = \frac{F(x_0 + \Delta x) - F(x_0 - \Delta x)}{2\Delta x}.$$

Since the nominal value is associated with the $u(x_0)$ solution, the variational expression is usually only evaluated for the upper and lower boundaries. This enables the evaluation of the range of variation of the static solution as

$$u_{max} = u(x_0 + \Delta x)$$

and

$$u_{min} = u(x_0 - \Delta x).$$

The number of derivatives to be computed may be raised if necessary, based on some magnitude criterion. It is not very expensive to go one derivative higher as the computation is recursive. On the other hand having larger number of design variable values could increase the computing time considerably.

An advancement over variational analysis is stochastic analysis. Latter is useful when a structural component exhibits a stochastic behavior. Such could occur, for example when material production procedures result in random variations of Young's modulus or density of the material.

One of the most popular stochastic approaches is the Monte-Carlo method. The method simulates the random response of the structure by computing deterministic responses at many randomly selected values of the design variable. After collecting these samples, the stochastic measures of the response variables, such as expected value or deviation, may be computed.

More advanced stochastic approaches use chaos theory based approximation of the response solution. Instead of giving a range to the design variable, it is considered to be a random variable with a certain distribution around a mean value

$$x \rightarrow (\xi, d(\xi)).$$

The stochastic solution is given by a polynomial expansion of form

$$\overline{u}(\xi) = \Sigma_{i=0}^{m} a_i \Psi_i(\xi),$$

where Ψ_i are basis functions chosen in accord with the type of the distribution function of the stochastic variable. For example, for normal distribution variable Hermite polynomials are used. The coefficients of the expansion are found by a least squares solution fitting above response to a set of known, deterministic solution points.

Both the Monte-Carlo and the chaos expansion solutions rely on a set of sampling solutions that are in themselves deterministic, hence these methods maybe executed externally to commercial finite element softwares. Finally, it is also possible to carry the stochastic phenomenon into the finite element formulation [2], however, this requires an internal modification to the finite element software.

References

[1] Gallagher, R. H.; Finite element analysis: Fundamentals, Prentice Hall, 1975

[2] Ghanem, R. G. and Spanos, P. D.; Stochastic finite elements: A spectral approach, Springer,, New York, 1991

[3] Hughes, T. J. R.; The finite element method, Prentice-Hall, Englewood Cliffs, New Jersey, 1987

[4] Jacobi, C. G., Über eines lechtes Verfahren die in der Theorie der seccularstörungen vorkommenden Gleichungen numerisch aufzulosen, Zeitschrift für Reine ungewandte Mathematik, Vol. 30, pp. 51-95, 1846

[5] Jankowic, M. S.; Exact nth derivatives of eigenvalues and eigenvectors, J. of Guidance and Control, Vol. 17, No. 1, 1994

[6] Nelson, R. B.; Simplified calculation of eigenvector derivatives, AIAA Journal, Vol. 14, pp. 1201-1205, 1976

[7] Vanderplaats, G. N.; An efficient feasible direction algorithm for design synthesis, AIAA Journal, Vol. 22, No. 11, 1984

18

Engineering Result Computations

After the solution computations are executed, the engineering solution set is needed. This process is sometimes called data recovery in commercial software [1]. The recovery operations are needed because the final solution may have been obtained in a reduced or in modal form, or some constraints may have been applied.

18.1 Displacement recovery

Some displacement recovery operations were already shown in the various chapters, we now collect them to illuminate the complete process.

If modal solution (Chapter 13) was executed, the analysis set solution is recovered by

$$u_a = \Phi_{ah} u_h,$$

where the Φ_{ah} is the modal space, assuming a frequency domain modal solution.

If, for example, dynamic reduction was executed (Chapter 11) then the recovery of the analysis set solution is executed by

$$u_a = S u_d,$$

where the S is the dynamic reduction transformation matrix.

Similar operations are needed when static condensation (Chapter 8) or component modal synthesis (Chapter 12) was executed. Of course, the analysis set solution may have been also obtained directly, in which case the prior steps are not executed.

The free set solution is obtained by

$$u_f = \begin{bmatrix} u_a \\ u_s^a \end{bmatrix},$$

where the u_s^a contains the automatically applied single-point constraints to remove singularities.

The recovery of the independent solution set, including the single-point constraints specified by the engineer, is done by

$$u_n = \begin{bmatrix} u_f \\ Y_s \end{bmatrix},$$

where the array Y_s contains the enforced displacements.

Finally, the engineering g partition of displacements is recovered by

$$u_g = \begin{bmatrix} I_{nn} \\ G_{mn} \end{bmatrix} u_n,$$

where the G_{mn} matrix represented the multi-point constraints applied to the system.

At this point, the displacement solution of the engineering model u_g is available. In some cases not all the components of the engineering solution set are obtained to execute this process expeditiously.

This is especially useful for example in optimization analyses, where some physical quantities are objective function components. Since the optimization process will be doing repeated analysis, in the intermediate stages only some components of the complete engineering solution set are needed.

Let us denote the partition of the analysis solution set interesting to the engineer by p and partition

$$u_a = \begin{bmatrix} u_p \\ u_{\bar{p}} \end{bmatrix},$$

where $\bar{p}$ is the remainder of the a partition. Then, for example the modal solution recovery may be significantly expedited in the form of

$$u_a = \Phi_{ph} u_h,$$

where

$$\Phi_{ah} = \begin{bmatrix} \Phi_{ph} \\ \Phi_{\bar{p}h} \end{bmatrix}.$$

Sometimes similar considerations are being used in partitioning the final results also. The g partition is also only computed in places where there is final result requested by the engineer. This, sometimes called sparse data recovery technology, could result in significant performance improvement on very large models in commercial environments.

18.2   Stress calculation

In mechanical system analysis, one of the most important issues besides the deformation (displacement solution) of the structure is the stress occurring in the structural components as that is the cornerstone of structural integrity. The subject of the analysis is to determine the deformations and related stresses under the given loads to simulate the real life work environment of the system.

Having computed all nodal displacements either in the time domain or in the frequency domain, the element nodal displacements q_e may be partitioned. It was established in Chapter 3 that

$$\epsilon = Bq_e$$

and

$$\sigma = D\epsilon.$$

Hence, the element stress is

$$\sigma = DBq_e,$$

where

$$\sigma = \begin{bmatrix} \sigma_x \\ \sigma_y \\ \sigma_z \\ \tau_{xy} \\ \tau_{yz} \\ \tau_{zx} \end{bmatrix}.$$

Here the σ are the normal and the τ are the shear stresses. They are most commonly evaluated at the element centroid. In practice the single stress formula of von Mises is often used. In terms of the normal and shear stresses it is written as

$$\sigma_v^2 = \frac{1}{2}[(\sigma_x - \sigma_y)^2 + (\sigma_y - \sigma_z)^2 + (\sigma_x - \sigma_z)^2 + 6(\tau_{xy}^2 + \tau_{yz}^2 + \tau_{zx}^2)].$$

This single stress value is computed for every element. In order to present the final results to the engineer these are converted to nodal values.

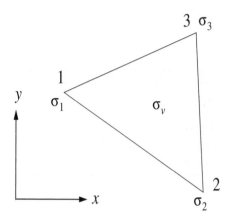

FIGURE 18.1 Stresses in triangle

18.3   Nodal data interpolation

In industrial practice it is common to calculate nodal values out of the constant element value with least square minimalization. Assume that the element value given is σ_v and we need corner values at a triangular element as shown in Figure 18.1.

With the help of the shape functions, the stress at a point in the element is expressed in terms of the nodal values as

$$\sigma = N_1\sigma_1 + N_2\sigma_2 + N_3\sigma_3,$$

where σ_i are the yet unknown corner stresses and the N_i are the shape functions as defined earlier.

Our desire is to minimize the following squared error for this element

$$E_e = \frac{1}{2}\int_A (\sigma_v - \sigma)^2 dA,$$

where σ_v is the computed element stress. Expanding the square results in

$$E_e = \frac{1}{2}\int_A \sigma_v^2 dA - \int_A \sigma_v \sigma dA + \frac{1}{2}\int_A \sigma^2 dA.$$

As σ_v is constant over the element, the first integral is simply a constant

$$\frac{1}{2}\sigma_v^2 A = C_e.$$

Introducing

$$\sigma_e = \begin{bmatrix} \sigma_1 \\ \sigma_2 \\ \sigma_3 \end{bmatrix},$$

the stress is the product of two vectors

$$\sigma = N\sigma_e,$$

where

$$N = \begin{bmatrix} N_1 & N_2 & N_3 \end{bmatrix}.$$

By substituting the second integral becomes

$$\sigma_v \int_A N\sigma_e dA = \sigma_e \sigma_v \int_A N dA,$$

as σ_e and σ_v now both are constant for the element. The third integral similarly changes to

$$\frac{1}{2}\sigma_e^T \int_A N^T N dA \; \sigma_e.$$

The evaluation of these element integrals proceeds along the same lines as the computations shown in Chapters 1 and 3. We sum (assemble) all the element errors as

$$E = \Sigma_{e=1}^m E_e = \Sigma_{e=1}^m (C_e - \sigma_v \sigma_e \int_A N dA + \frac{1}{2}\sigma_e^T \int_A N^T N dA \sigma_e),$$

where m is the number of elements in the finite element model. We then introduce

$$S = \Sigma_{e=1}^m \sigma_e = \begin{bmatrix} \sigma_1 \\ \sigma_2 \\ \sigma_3 \\ \cdots \\ \sigma_{n-2} \\ \sigma_{n-1} \\ \sigma_n \end{bmatrix},$$

where n is the number of nodes in the finite element model. Also introduce

$$R = \Sigma_{e=1}^m \int_A N^T N \, dA,$$

and

$$T = \Sigma_{e=1}^m \sigma_v \int_A N \, dA.$$

With these the assembled error is

$$E = \frac{1}{2} S^T R S - S^T T + C.$$

We obtain the minimum of this error when

$$\frac{\partial E}{\partial S} = 0,$$

which occurs with

$$RS = T,$$

since $C = \Sigma_{e=1}^m C_e$ is constant. Note, that the calculation of R is essentially the same as the calculation of the mass matrix in Section 3.5, apart from the absence of the density ρ. Hence, the matrix of this linear system is similar to the mass matrix in structure and as such very likely extremely sparse or strongly banded. Therefore, this linear system may now be solved with the techniques shown in Chapter 7. The resulting nodal stress values in S form the basis for the last computational technique discussed in the next section.

18.4   Level curve computation

The nodal values, whether they were direct solution results such as the displacements or were interpolated as the stresses, are best presented to the engineer in the form of level, or commonly called, contour curves.

Let us consider the example of the triangular element with nodes $1, 2, 3$ and nodal values $\sigma_1, \sigma_2, \sigma_3$ shown in Figure 18.2.

We are to find the location of the points R, S on the sides of the triangle that correspond to points P, Q that define a constant level of stress. Let the value of the level curve be σ_l. Introduce the ratio

$$p = \frac{\sigma_l - \sigma_1}{\sigma_3 - \sigma_1}.$$

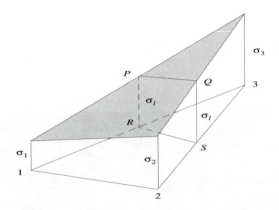

FIGURE 18.2 Level curve computation

Some arithmetic yields the coordinates of point R as

$$x_R = px_3 + (1-p)x_1,$$

and

$$y_R = py_3 + (1-p)y_1.$$

Similarly the ratio

$$r = \frac{\sigma_l - \sigma_2}{\sigma_3 - \sigma_2}$$

yields the coordinates of S as

$$x_S = rx_3 + (1-r)x_2,$$

and

$$y_S = ry_3 + (1-r)y_2.$$

These calculations, executed for all the elements, result in continuous level curves across elements and throughout the finite element model.

It should be noted that above method is fine when the stresses of the exact solution are smooth. The method could result in large errors in cases when the exact stress is not smooth, for example, in elements located on singular edges of the model.

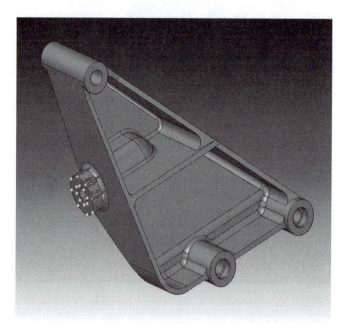

FIGURE 18.3 Physical load on bracket

18.5 Engineering analysis case study

The engineering analysis process and the computed results are demonstrated by the bracket model for which the geometric and finite element models were presented in Chapter 2. The first step is to apply the loads to the model as shown on Figure 18.3 by a pressure load applied horizontally to the left face of the model.

FIGURE 18.4 Constraint conditions of bracket

Note, that in modern engineering environments, such as the NX CAE environment of the virtual product development suite of Siemens PLM Software [2], the loads are applied to the geometric model in accordance with the physical intentions of the engineer. In this case the load is distributed on the face bounded by the highlighted edges and the direction of the load is denoted by the arrows.

The constraints are also applied to the geometric model. In the case of the example the constraints were applied to the interior cylindrical holes where the bolts will be located and made visible by the arrows on Figure 18.4.

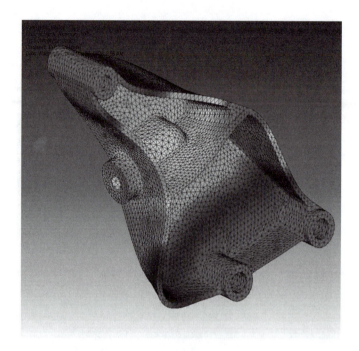

FIGURE 18.5 Deformed shape of bracket

The computational steps of the analysis process were executed by NX NAS-TRAN [3]. The displacements superimposed on the finite element mesh of the model result in the deformed shape of the model as shown in Figure 18-5. This is one of the most useful informations for the engineers.

The stress results projected to the undeformed finite element model are visualized in Figure 18.6. The lighter shades indicate the areas of higher stresses. The level curve representation of the results described in the last section was used for the contours.

The NX environment also enable the animation of the deformed shape with a contiguous process between the undeformed and the deformed geometry, another useful tool for the engineer.

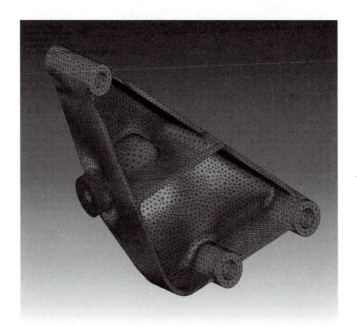

FIGURE 18.6 Stress contours of bracket

References

[1] Craig, R. R. Jr.; Structural dynamics, An introduction to computer methods, Wiley, New York, 1981

[2] www.plm.automation.siemens.com/en_us/products/nx/design/index.shtml

[3] www.siemens.com/plm/nxnastran

Annotation

Notation	Meaning
P	Potential energy, permutation
P_k	Householder matrix
T	Kinetic energy, transformation matrix
D	Dissipative function, material matrix
D	Diagonal factor matrix
W_P	Work potential
P_s	Strain energy
N_i	Matrix of shape functions
N	Shape functions
B	Strain displacement matrix
J	Jacobian matrix
q_e	Element displacement
k_e	Element stiffness
A_e	Element area
V_e	Element volume
E_e	Element energy
K	Stiffness matrix
M	Mass matrix
B	Damping matrix
F	Force matrix
G	Static condensation matrix
H^1	Hilbert space
I	Identity matrix
C	Cholesky factor
L	Lower triangular factor matrix
U	Upper triangular factor matrix
S	Dynamic reduction transformation matrix
S_j	Spline segment
P_i	Point coordinates
R	Rigid constraint matrix
$V(P_i)$	Voronoi polygon
Y_s	Vector of enforced displacements
T	Tridiagonal matrix
Q	Permutation matrix, Lanczos vector matrix

X	Linear system solution
Y	Intermediate solution
Z	Residual flexibility matrix
ϵ	Strain vector
σ	Stress vector
α	Diagonal Lanczos coefficient
β	Off-diagonal Lanczos coefficient
λ	Eigenvalue, Lagrange multiplier
Λ	Eigenvalue matrix
ϕ	Eigenvector
Φ	Eigenvector matrix
Ψ	Residual flexibility matrix
ω	Frequency
μ	Shifted eigenvalue
λ_s	Spectral shift
Δt	Time step
ΔF	Nonlinear force imbalance
Δu	Nonlinear displacement increment
κ_k	Krylov subspace
θ	Rotational degrees of freedom
b_i	Modal damping
$f(x)$	Objective function
$g(x)$	Constraint function
k_i	Modal stiffness
m_i	Modal mass
q_k	Lanczos vectors
q_i	Generalized degrees of freedom
r	Residual vector
t	Time
u	Displacement in frequency domain
v	Displacement in time domain
w	Modal displacement
w_i	Weight coefficients

List of Figures

List of Tables

Index

acoustic
  response, 113
active forces, 61
algebraic equation of motion, 71
axial bar, 77

backward whirl, 194
basis functions, 9
BFGS method, 282
biconjugate gradient method, 136
block
  method, 164
boundary conditions, 85
boundary mode shapes, 202
boundary partition, 141, 201
brake
  automobile, 196
buckling analysis, 72

CAD, 54
Campbell diagram, 194
central difference method, 251, 252,
  256
Closing Remarks, 331
complex eigenvalues, 183, 197
component mode synthesis, 217
condensation matrix, 142
congruence transformation, 167, 207,
  226, 227
conjugate gradient method, 135
constraint matrix, 78, 213
constraint sensitivities, 291
Coriolis matrix, 192
critical speed, 194

damping matrix, 71, 72, 183
degrees of freedom, 61, 75, 80, 86,
  97

dependent degrees of freedom, 78,
  80
design
  objective, 291
  optimization, 290
  sensitivity, 289
  variables, 289
displacements, 62
dissipative forces, 61
dynamic reduction, 201, 202

eigenvalue sensitivity, 302
eigenvalues, 166, 185
  closely spaced, 159
  distribution of, 161
  multiple, 168
  physical, 186, 188
  recovery of, 164
eigenvector sensitivity, 303
eigenvectors, 162, 164, 166, 167, 185
  left, 190
  mathematical, 190
  physical, 188, 189
enforced displacements, 86
equilibrium equation, 71
error
  approximated, 163, 168
  bound, 162
  norm, 186
  residual, 163, 168

factorization, 120, 161, 167
  symmetric, 160
feasible direction, 292
finite differences, 257, 290, 291, 303,
  306
finite element mesh, 12

Closing Remarks

The book's goal was to give a working knowledge of the main computational techniques of finite element analysis. Extra effort was made to make the material accessible with the usual engineering mathematical tools.

Some chapters contained a simple computational example to demonstrate the details of the computation. It was hoped that those help the reader to develop a good understanding of these computational steps.

Some chapters contained a description of an industrial application or an actual real life case study. They were meant to demonstrate the awesome practical power of the technology discussed.

In order to produce a logically contiguous material and to increase the readability, some of the more tedious details were omitted. In these areas the reader is encouraged to follow the cited references.

The reference sections at the end of each chapter are organized in alphabetic order by authors' names. They are all publicly available references. Many original publications on topics contained in the book are given. The best review references are also cited, especially those dealing with the mathematical, engineering and geometric theory of finite elements.